www.KnowledgePublications.com

SAVE HARMLESS AGREEMENT

Because use of the information, instructions and materials discussed and shown in this book, document, electronic publication or other form of media is beyond our control, the purchaser or user agrees, without reservation to save Knowledge Publications Corporation, its agents, distributors, resellers, consultants, promoters, advisers or employees harmless from any and all claims, demands, actions, debts, liabilities, judgments, costs and attorney's fees arising out of, claimed on account of, or in any manner predicated upon loss of or damage to property, and injuries to or the death of any and all persons whatsoever, occurring in connection with or in any way incidental to or arising out of the purchase, sale, viewing, transfer of information and/or use of any and all property or information or in any manner caused or contributed to by the purchaser or the user or the viewer, their agents, servants, pets or employees, while in, upon, or about the sale or viewing or transfer of knowledge or use site on which the property is sold, offered for sale, or where the property or materials are used or viewed, discussed, communicated, or while going to or departing from such areas.

Laboratory work, scientific experiment, working with hydrogen, high temperatures, combustion gases as well as general chemistry with acids, bases and reactions and/or pressure vessels can be EXTREMELY DANGEROUS to use and possess or to be in the general vicinity of. To experiment with such methods and materials should be done ONLY by qualified and knowledgeable persons well versed and equipped to fabricate, handle, use and or store such materials. Inexperienced persons should first enlist the help of an experienced chemist, scientist or engineer before any activity thereof with such chemicals, methods and knowledge discussed in this media and other material distributed by KnowledgePublications Corporation or its agents. Be sure you know the laws, regulations and codes, local, county, state and federal, regarding the experimentation, construction and or use and storage of any equipment and or chemicals BEFORE you start. Safety must be practiced at all times. Users accept full responsibility and any and all liabilities associated in any way with the purchase and or use and viewing and communications of knowledge, information, methods and materials in this media.

SERI/SP–451–519
UC–61

FUEL FROM FARMS

A Guide to Small-Scale
Ethanol Production

A Product of the
Solar Energy Information Data Bank

Solar Energy Research Institute
Operated for the U.S. Department of Energy by the
Midwest Research Institute

Under Contract No. EG–77–C–01–4042

February 1980

TABLE OF CONTENTS

LIST OF TABLES AND FIGURES

ACKNOWLEDGMENTS

Mr. Ted D. Tarr, of the U.S. Department of Energy's Division of Distributed Solar Technology, has provided overall policy guidance and direction in the development of *Fuel from Farms — A Guide to Small-Scale Ethanol Production*.

The Solar Energy Research Institute's Solar Energy Information Data Bank (SEIDB) staff was requested to prepare on a quick-return basis this guide on the small-scale production and use of fermentation ethanol. The effort at SEIDB was directed by Mr. Paul Notari and coordinated by Mr. Stephen Rubin of the Information Dissemination Branch.

The team created by SEIDB represented a cross-section of knowledgeable individuals and consultants familiar with small-scale fermentation ethanol techniques. The team consisted of the following individuals:

Project Manager	**Project Technical Director**
Mr. V. Daniel Hunt	Mr. Steven J. Winston
TRW Energy Systems Group	Energy Incorporated

Consultants

Dr. Billy R. Allen	Mr. Pincas Jawetz
Battelle Columbus Laboratories	Consultant on Energy Policy
Mr. Jerry Allsup	Mr. Donald M. LaRue
Bartlesville Energy Technology Center	EG & G Idaho, Inc.
Dr. Mani Balasubramaniam	Mr. Robert A. Meskunas
TRW Energy Systems Group	Environmental Group
Mr. William P. Corcoran	Dr. Thomas Reed
Solar Energy Research Institute	Solar Energy Research Institute
Mr. David Freedman	Mr. Jim Smrcka
Center for the Biology of Natural Systems	Galusha, Higgins and Galusha
Mr. William S. Hedrick	Dr. Ruxton Villet
Consulting Engineer	Solar Energy Research Institute
Mr. Jack Hershey	Dr. Harlan L. Watson
Environmental Group	TRW Energy Systems Group

Editorial and Production Support

Our appreciation is extended to

- production editor Doann Houghton-Alico for her careful attention to detail and in-depth editing of this book;

- Rick Adcock, special editor, and assistant editors Elzene Sloss, Holly Wilson, and Diane Schilly;

- graphic designers Raymond David, Lianne Mehlbach, and Laura Ehrlich;

- the production staff: Linda Arnold, Angela Bagley, Linda Baldwin, Delores Barrier, Pat Bledsoe, Donna Clausen, Elaine Davis, Judy Davis, Penelope Eads, Donna Esile, Pat Haefele, Meribah Henry, Diane Howard, Francis Langford, Agnes Mdala, Nghia Pham, Marylee Phillips, Walt Shipman, Kathy Sutton, Susie Van Horn, and Ruth Yeaman;

- the printing staff: Vicki Doren, Mike Linenberger, and Donna Miller.

Photographs

The agricultural photographs are reprinted through the courtesy of Grant Heilman Photography. The photograph on page 9 appears courtesy of the Iowa Development Commission.

Reviewers

We acknowledge the following individuals for their helpful reviews of the draft of *Fuel from Farms — A Guide to Small-Scale Ethanol Production*. These individuals do not necessarily approve, disapprove, or endorse the report for which SEIDB assumes responsibility.

Ms. Daryl Bladen
U.S. Department of Commerce

Mr. Ken Bogar
Montana Farmer

Mr. Robert E. Brown
Ohio Farm Bureau Federation

Mr. David L. Feasby
Solar Energy Research Institute

Mr. Charlie Garlow
Public Interest Research Group

Mr. Denis Hayes
Solar Energy Research Institute

Mr. Jack Hill
Ohio Farm Bureau Federation

Mr. William C. Holmberg
U.S. Department of Energy

Dr. T. Q. Hutchinson
U.S. Department of Agriculture

Mr. Edward A. Kirchner
Davy McKee Corp., Chicago

Dr. Michael R. Ladisch
Purdue University

Mr. Herbert Landau
Solar Energy Research Institute

Dr. Leslie S. Levine
U.S. Department of Energy

Dr. Edward S. Lipinsky
Battelle Columbus Laboratories

Mr. Irving Margeloff
Publicker Industries

Dr. Paul Middaugh
South Dakota State University

Mr. Strud Nash
E. F. Hutton & Company

Mr. Michael H. Pete
Consultant
U.S. Department of Energy

Dr. Bruce L. Peterson
American Petroleum Institute

Mr. E. Stephen Potts
U.S. Department of Energy

Mr. Myron Reamon
National Gasohol Commission

Dr. Clayton Smith
Solar Energy Research Institute

Dr. Jon M. Veigel
Solar Energy Research Institute

This document greatly benefited from the many previous efforts in alcohol fuels including working groups at universities and colleges (such as Colby Community College), private marketing efforts, private research and development projects, and individual efforts to collect and organize information pertinent to alcohol fuels.

Introduction

CHAPTER I
Introduction

OBJECTIVE

The expanding support for gasohol in this country over the last several years provides an opportunity to directly reduce U.S. oil imports in the very near future. Interest is evident by the many requests for information about gasohol that are being received throughout the federal government daily. This guide has been prepared to meet the challenge of filling the information void on fermentation ethanol in a balanced, reasoned way, with emphasis on small-scale production of fermentation ethanol using farm crops as the source of raw materials. It is addressed not only to those in the U.S. farming community who may wish to consider the production of ethanol as part of their normal farming operations, but also to owners of small businesses, investors, and entrepreneurs.

This guide presents the current status of on-farm fermentation ethanol production as well as an overview of some of the technical and economic factors. Tools such as decision and planning worksheets and a sample business plan for use in exploring whether or not to go into ethanol production are given. Specifics in production including information on the raw materials, system components, and operational requirements are also provided. Recommendation of any particular process is deliberately avoided because the choice must be tailored to the needs and resources of each individual producer. The emphasis is on providing the facts necessary to make informed judgments.

PERSPECTIVE

Foreign crude oil imports currently provide the raw material for production of half of the liquid fuels consumed in the United States and represent a cash outflow of almost $8 million per hour. Recent events have dramatically illustrated the substantial economic cost, instability, and economic vulnerability of such imports. Ethanol is a liquid fuel that can substitute domestic renewable resources for petroleum products now and increasingly in the next few years.

Fermentation ethanol is becoming the first nonpetroleum fuel to attain widespread use in the United States. This trend is apparent from the rapid increase in the sale of gasohol, a blend of 10% agriculturally derived anhydrous ethanol and 90% unleaded gasoline. As of late 1979, the market had expanded to more than 2,000 outlets in 35 states. Gasohol can be readily substituted for unleaded gasoline in current vehicles with no engine adjustments and little or no change in engine performance.

The petrochemical market for fermentation ethanol, while considerably smaller than the automotive fuel market, is also substantial. Thirty percent of the bulk of industrial-grade ethanol is produced from a petroleum derivative and hence is also a potential candidate for displacement by fermentation ethanol.

The production of ethanol from grain leaves behind a protein-rich stillage. This stillage, used in conjunction with straw, permits reduction in the use of hay and grain, and becomes an excellent, nutritive source of animal feed. Dried stillage, in turn, can also be exported as feed with practically no loss in commercial value.

The supply of ethanol is still limited. Essentially, all of the ethanol used in gasohol is currently obtained from a few producers, in spite of the expanding market. However, the production intended for automotive uses is increasing. In early 1979, production of ethanol for gasohol was at a rate of 30 million gallons annually. It is expected that by the end of 1980 this will increase quite significantly.

Existing and proposed federal and state incentives for fermentation ethanol production and use have contributed to the rapid expansion of the gasohol market. In addition, a broad spectrum of options is currently being pursued at the federal level to help accelerate the commercialization of gasohol by stimulating both its production and uses. Maximizing ethanol production will require a mix of various sized ethanol plants. Because of the lag time involved in building and operating larger facilities, it is critical to provide basic information to individuals interested in constructing small-scale facilities—since they can be built most quickly.

The production of fermentation ethanol is based on established technology, and a variety of raw materials is available from the agricultural sector to more than meet projected demands. Fermentation ethanol can be produced from such crops as corn, wheat, sugarcane,

Small-Scale Ethanol Production can be Readily Incorporated into Farm Operations

potatoes, cassava, beets, and Jerusalem artichokes; from agricultural byproducts and wastes; and from cellulose. In short, whatever can be broken down to sugars can become a primary material for fermentation. Thus, the variety of raw materials is quite large. These crops as well as distressed grains are ideal for the production of fermentation ethanol and do not affect the availability of food supplies.

The United States has the potential for growing grains and other crops well in excess of the requirements for domestic and export markets. Economic factors have consequently played a major role in the institution of "set-aside land" and "land diversion" programs by the U.S. Department of Agriculture (USDA). However, growing grain or other crops on this land for fuel production would not detract from the production of food. Rather, if properly utilized, it would constitute a resource that would otherwise have been left idle. Furthermore, the crops grown on this land can still be held in reserve for emergency food, should that become necessary. In 1978, for example, the USDA has certified that the amount of cropland left fallow was 13.4 million acres under the set-aside program and an additional 5.3 million acres under the diversion program. If this acreage had been cultivated with corn for ethanol production, nearly 3.03 billion gallons of ethanol and 10 million tons of distillers' dried grains (DDG) could have been produced. (This assumes a modest average yield of 65 bushels of corn per acre per year with an average production of 2.5 gallons of 200-proof ethanol and 17 pounds of DDG per bushel of corn.) This is only ethanol produced from land left idle through two specific farm programs. The production of fermentation ethanol is not limited by the extent of this land, and additional unused land as well as some land currently under cultivation can be used for crops for production of fermentation ethanol. All this makes the production of ethanol even more promising, and a conservative estimate for the potential displacement of petroleum is at least several billion gallons per year in the near term.

Belt tightening alone will not help the United States solve the present economic difficulties. Farmers, like everyone else, do not like austerity programs and would rather increase our national wealth. This can be achieved by increasing productivity—the production of more goods and services from every barrel of oil we use and development of new sources of energy.

Clearly, the agricultural sector has a role whose full potential is just beginning to be realized. A farm-based fermentation ethanol industry can provide a decentralized system of fuel production and a measure of energy self-sufficiency for the farm community. This can be accomplished as an integral part of normal farming operations following sound agricultural practices.

The technology for ethanol production has existed for centuries. In the early 1900's, Henry Ford and others in the U.S. auto industry used ethanol as the fuel for automobiles. Ultimately, it was replaced by gasoline, which was much cheaper. Today, the tables appear to be turning once again, this time in favor of fuel derived from renewable domestic resources. There are, however, several underlying issues related to fermentation ethanol production that must be examined.

ISSUES

In addition to the need to increase the number of ethanol production facilities, there is the concern about the impact of ethanol production on agriculture. The production of more ethanol than is obtainable from surplus and distressed crops will require cultivation of land that is currently fallow and shifts to specialized high-yield crops. The switch to such crops may allow a decrease in use of fertilizers, pesticides, and herbicides, whose production and transport require petroleum fuels and natural gas. This diversification of crops itself offers specific advantages to the farmer, not least of which may be modifications of agricultural practice and new patterns of crop rotation to improve soil fertility. Nongrain forage crops need less fertilizer, herbicides, and pesticides than high-yield grain crops. As commercial processes become available for the small-scale conversion of these crops to ethanol, the opportunity will exist to decrease demands on the soil to achieve production of equivalent value to current crops.

The energy balance of ethanol production and use is a controversial subject. Whether one achieves a net energy gain or loss in ethanol production depends upon where the energy boundaries are drawn and the assumptions used. Examples of alternative means of determining the energy balance in ethanol production are given in Appendix D.

Conversion of crops with significant human food value to fuel is not desirable. Fortunately, production of fermentation ethanol does not make this an "either-or" consideration. Much of the cereal grain (including most of the corn) currently produced in the United States is used as animal feed. While fermentation of cereal grains to produce ethanol uses most of the carbohydrates, almost all of the protein is recovered in the stillage coproduct. This stillage can be fed directly to animals as a high-protein source, and other nutritional requirements can be filled using forages which have no value as human food. This consideration, along with the use of spoiled perishable crops, distressed crops, and marginal crops, provides a feedstock base for ethanol production that requires no displacement of crops for human food.

Since stillage is considered an animal feed replacement for soybean meal (on a protein equivalence basis), there is legitimate concern about its impact on the soybean meal market. However, this concern has to be viewed in the proper perspective. First, the use of soybean meal and cotton seed meal for animal feed was developed after World War II. Their use changed the entire animal feed pattern in the United States and, in the process, displaced grains such as corn, oats, wheat, barley, and high-quality hay. Second, from the general viewpoint of the farm community, agricultural products must be able to compete for markets on an equal footing. Consequently, if stillage proves to be economically and nutritively more attractive than soybean meal, markets for it must be allowed to develop normally. One can thus predict a healthy readjustment of farm production to a new set of conditions that will develop with the introduction of fermentation ethanol.

Another issue is anhydrous ethanol versus hydrated ethanol production. Anhydrous ethanol is more costly and energy intensive to produce than lower proof ethanol. However, if the ethanol is to be sold to blenders for use in gasohol, the ability to produce

There is Sufficient Land Available to Produce the Quantity of Crops Needed to Achieve Ethanol Production Goals

Small-Scale Ethanol Production can Take Advantage of the Types of Crop Storage and Handling Equipment Already in use on the Farm

anhydrous ethanol is mandatory. Hydrated ethanol may be produced for on-farm use or for use in topping cycles.

A final but important issue of concern is the economics of small- and large-scale production of fermentation ethanol. In most production processes, substantial economies-of-scale are realized with higher plant capacity. However, in the case of on-farm fermentation ethanol production, certain economies-of-scale are also present for small-scale production (e.g., lower transportation and capital costs) which may balance the economic advantages of large-scale plants. As a result, small-scale production of ethanol may possibly be achieved with product costs comparable to those from larger plants. Thus, there appears to be a future role for both small- and large-scale plants for the production of fermentation ethanol.

GUIDE TO THE DOCUMENT

A detailed consideration of the several factors briefly discussed above is presented in the six chapters and appendices that follow.

A decision process to determine the feasibility of on-farm production of ethanol is developed in Chapter II, with emphasis on the market for ethanol and what must

be done to participate in it profitably. The sequential steps involved in this process are presented in planning and decision worksheets.

Ethanol production operations are described in Chapter III to indicate how the conversion of agricultural products proceeds through the various stages. Feedstock considerations are discussed in Chapter IV with particular attention to alternate crops, their ethanol yield potentials, and overall implications of their respective agricultural requirements. Ethanol plant design considerations are treated in detail in Chapter V. They include (1) farm-related objectives and integration of ethanol production with normal farming operations; (2) plant design criteria and functional specifications; and (3) energy, labor needs, process control, and safety aspects, and the inherent tradeoffs between them. The information developed is then applied to the design of a representative, small-scale fermentation ethanol production plant, with an output of 25 gallons of anhydrous ethanol per hour. All major operational features are addressed, including the requirements for system control, record keeping, and maintenance. This representative plant is intended to serve as a model from which an actual facility can be designed, built, and operated.

Chapter VI follows with a detailed preparation of a business plan for building the 25-gallon-per-hour facility. The business plan draws on information developed in Chapter V. Its purpose is to determine the financial obligations of the farm owner and the profitability of the enterprise, both of which are essential to obtain necessary financing for construction and operation. Alternative sources of financing available to the small-scale farmer are described and their special requirements are identified. As in Chapter V, the material in Chapter VI is intended to serve as a basis from which an actual business plan can be prepared.

The appendices complete the handbook. They provide a description of current regulations and legislation at the federal and state levels concerning fermentation ethanol production; information on plant licensing and bonding requirements enforced by the Bureau of Alcohol, Tobacco, and Firearms; discussion of the environmental considerations that apply to on-farm production of ethanol; reference data and charts; lists of resources, both people and information; a bibliography; and a glossary.

CHAPTER II
Decision to Produce

CHAPTER II
Decision To Produce

Expanding farm operations to include a fermentation ethanol plant is ultimately a personal decision. Information can be collected and planning tools used to provide a foundation for such a decision. Market uses and assessment for fuel ethanol and stillage as coproducts, production potential, equipment selection, and financial requirements are the four major areas to be considered in this chapter, which, with succeeding chapters as building blocks, is intended to set up the tools for the decision-making process.

Market values can be estimated for all the products and used as a basis for evaluating the profit potential, which can then be examined in relation to the complete farm operation. Direct considerations affecting production potential, such as how much feedstocks can be grown and how much ethanol can be produced, are also examined. The decision and planning worksheets at the end of this chapter can be used as a step-by-step tool for reaching a decision on whether or not to develop a small-scale, on-farm fermentation ethanol plant.

In addition to the direct factors examined in the worksheets there will be intangible considerations, such as a desire for on-farm fuel self-sufficiency.

With Proper Modification, Straight Ethanol can be Used in Either Gasoline- or Diesel-Powered Farm Equipment

BENEFITS

There are three areas in which there are benefits to the farm economy from small-scale, on-farm, ethanol production. These are direct sales, on-farm uses, and indirect farm benefits.

Farm-produced ethanol sold for profit provides an alternative market for farm commodities. It can provide a "shock absorber" for excess production and a "fall back position" if unforeseen events adversely affect crop or yields.

Farming, perhaps more than any other single occupation, offers the opportunity for self-reliance. The on-farm production of ethanol expands this opportunity. Ethanol can be used in farm equipment as a blend with gasoline in spark ignition engines, as anhydrous or hydrated ethanol fuels in modified spark ignition engines, as a blend with diesel fuel in diesel engines, and as a dual-carbureted mixture with water in diesel turbochargers to enhance efficiency. Protein coproducts, such as stillage, can be fed to farm animals as

a replacement for other protein sources. Cellulosic coproducts, if sufficiently dry, can be burned as fuel.

Farm overproduction is generally planned to meet anticipated demand in the event of possible reductions in crop yield. However, the cumulative result of consistent overproduction in the absence of alternative markets is depressed commodity prices. Consequently, the financial health of many farms depends on the opening of new markets. Fermentation ethanol production provides several alternative markets for a broad variety of farm commodities.

MARKETS AND USES

Ethanol

The use of ethanol for fuel in internal combustion engines is not a new concept. Engines built around the turn of the century used ethanol for fuel. Henry Ford offered automobiles capable of operating on either ethanol or gasoline [1]. With the development of equipment capable of economically extracting and refining petroleum early in this century, gasoline became the more practical fuel and further development of fuel-grade ethanol was shelved. Now that the production from domestic petroleum reserves is becoming more costly and difficult to develop and foreign oil is at a premium, the nation is looking for ethanol to displace some petroleum-based fuels and chemicals.

The forms in which ethanol can be used for fuel are as

- various ethanol-gasoline blends,

- hydrated (lower proof) ethanol,

- straight anhydrous ethanol, and

- dual-carbureted diesel fuel supplement.

Anhydrous Ethanol can be Blended With Gasoline for Direct Use in Unmodified Vehicles

Ethanol-Gasoline Blends. The market for gasohol (a blend of 90% unleaded gasoline and 10% agriculturally derived anhydrous ethanol) is already well established in many parts of the country. It is expected that by the end of 1981 as much as 500 million gallons of ethanol production capacity could be made available to make gasohol, and that within 5 years this quantity could increase three to ten times.

Hydrated Ethanol. Hydrated ethanol can be burned efficiently in spark-ignited internal combustion engines with minor alterations to the engine. Regular motor vehicle engines have been successfully modified to run on to ethanol. The jet size in the carburetor needs to be enlarged slightly when converting from a gasoline to an ethanol-powered engine because ethanol contains less useful thermal energy per unit volume than gasoline. Accordingly, more ethanol than gasoline must be introduced into an engine to generate the equivalent amount of energy. With most engines, it is also necessary to modify the intake manifold to insure proper vaporization of the ethanol so that all cylinders will be operating with equal air-fuel mixtures. There are many possible methods for doing this, such as installing preheaters in the fuel system or enlarging the heat stove on the exhaust manifold, with accompanying adjustment of the heat stove control gate for the higher temperature requirement. However, none of these systems are commercially available. However, problems associated with the burning characteristics of the ethanol-water mixture can complicate performance and become a serious impairment as the concentration of water increases.

Anhydrous Ethanol. Anhydrous ethanol can be burned directly in spark-ignition engines using essentially the same engine modifications discussed above for the use of hydrated ethanol. However, hydrated ethanol is less costly and it is not likely that anhydrous ethanol would

find extensive use as motor fuel. Its primary use is likely to be as an additive to gasoline to produce gasohol.

In the United States, gasohol usage has been demonstrated in a large number of tests to be a motor fuel essentially equivalent to gasoline. Gasohol does have less total thermal energy per unit volume than gasoline, however, no significant decrease in terms of "miles per gallon" results from the use of gasohol.

The addition of ethanol to gasoline increases the octane rating of the mixture because anhydrous ethanol is a high-octane fuel. In the past, the octane of fuels was increased by adding tetraethyllead. Because the lead compounds have significant adverse impacts on the environment, the conversion to unleaded gasoline was mandated. The changes in refinery operations that are required to produce fuel of the same octane without lead reduce the quantity of fuel that can be produced from a barrel of crude oil. This is because the chemical constituency of the gasoline is altered by reforming lower hydrocarbons to increase the percentage of octane-boosting aromatic compounds. This reforming process consumes additional energy in the refining process— energy directly lost from every barrel processed. The addition of ethanol to gasoline effectively gives the required octane boost and the reforming requirement is correspondingly reduced. This means that every barrel of gasohol produced decreases crude oil demand not only by the quantity of gasoline directly replaced by ethanol, but also by the crude oil saved due to the value of ethanol as an octane enhancer [2].

The use of a mixture of hydrated ethanol and unleaded gasoline can lead to complications. Mixtures of water, ethanol, and gasoline can encounter problems when the three components do not remain in solution. Depending upon the amount of water, the characteristics of the gasoline, and the temperature, two distinct phases can

Corn Stover can be Burned to Provide the Heat Needed for Ethanol Production

separate out. When this separation occurs, the upper phase (layer) is comprised of gasoline and the lower phase (layer) is comprised of water and most of the ethanol. Because the air-fuel requirements are different for the ethanol and gasoline fuels, vehicle operations will not be satisfactory if the fuels separate. Heating and agitating these two phases will cause them to go back into solution, but subsequent cooling will result in phase separation again.

Data from tests on gasohol used in vehicles in Brazil and domestically in Nebraska, Iowa, Indiana, and other states indicate no adverse effects on engine life.

Dual-Carbureted Diesel Fuel Supplement. Diesel engines can operate on separately carbureted ethanol and diesel fuel. When low-quality diesel fuel is used, the amount of ethanol injected is generally less than 25%. When the intent is to reduce "diesel smoke" and increase power, the amount of ethanol used can range as high as 50% [3].

Industrial Chemical Feedstocks. The chemical industry consumes large quantities of ethanol either as a basic feedstock or for use as a solvent. Most of the ethanol currently used by industry is produced from petroleum- or natural gas-derived ethylene. Thus, the cost of ethylene conversion to ethanol is a direct function of petroleum and natural gas costs. As petroleum-derived ethanol costs continue to increase, industrial consumers will look for less expensive sources of ethanol. The current selling price of ethanol produced in this manner is in excess of $1.50 per gallon [4]. These markets are highly localized and generally far removed from rural areas.

The largest industrial chemical markets in the United States are for acetic acid and ethylene because of their wide use in the production of polymers. Acetates (acetic acid polymers) constitute the raw material for synthetic fabrics, plastics, and an enormous variety of common products. Ethanol can be fermented directly to acetic acid (this is what happens when wine turns to vinegar). Acetic acid is also a byproduct of ethanol fermentation. Hence, consideration may be given to recovery of this material.

The pharmaceutical industry also consumes large quantities of ethanol for use as solvent. The quality control requirements for this market are extremely stringent and the costs of producing a pure product (not just anhydrous, but free of fusel oils and other contaminants) is quite high.

Fermentation ethanol has replaced a significant portion of petroleum-derived ethanol in India and Brazil [5, 6]. In fact, ethylene is produced from fermentation ethanol in these countries. Similar programs are being developed in the Philippines, South Africa, Australia, and other countries, and it is reasonable to assume that such a development could also occur in the United States.

Other Uses. Other possible uses of ethanol are as fuel for

• crop drying,

• general heating, and

• electricity generation with small generators.

Coproducts

Stillage can be fed to farm animals as a protein supplement either whole (as produced), wet, solid (screened), or dry. The stillage from cereal grains ranges from 26% to 32% protein on a dry basis. The basic limitation on the amount that can be fed at any one time to an animal is palatability (acid concentration caused by drying makes the taste very acrid). Mature cattle can consume about 7 pounds of dry stillage per day or, roughly, the stillage resulting from the production of 1 gallon of ethanol. The feeding of whole stillage is limited by the normal daily water intake of the animal and the requirements for metabolizable energy and forage fiber. The feeding value to swine and poultry is somewhat limited. Wet stillage cannot be stored for long periods of time, and the lack of locally available herds of animals to consume it may lower its value. Stillage from grains contaminated with aflatoxins cannot be used as animal feed.

The cellulosic coproducts may be directly fermented to produce methane gas or dried for use as boiler fuel.

Carbon dioxide (CO_2) produced by fermentation can be compressed and sold to users of refrigerants, soft drink bottlers, and others. It also has many agricultural applications which are beyond the scope of this handbook.

MARKET ASSESSMENT

Before a decision to produce can be made, it is necessary to accurately determine if markets for the ethanol and coproducts exist close enough to allow for economical distribution. The size of the market is defined by the quantities of ethanol and coproducts that can be used directly on the farm and/or sold. The ethanol on-farm use potential can be determined from the consumption of gasoline and diesel fuel in current farming operations. Then, a decision must be made on the degree of modification that is acceptable for farm equipment. If none is acceptable, the on-farm use will range from 10% to 20% of the total gasoline consumption. If direct modification to spark ignition equipment is acceptable, the on-farm potential use can be 110% to 120% of current gasoline consumption. If the risks associated with attempting undemonstrated technology are considered acceptable, the ethanol replacement of diesel fuel will be roughly 50% of current diesel fuel consumption [3].

The sale of ethanol off the farm will be dependent upon local conditions and upon the type of Bureau of Alcohol, Tobacco, and Firearms (BATF) license obtained. (Currently, a commercial license from BATF is required for off-farm sale of ethanol.) Market estimates should be based on actual letters of intent to purchase, not an intuitive guess of local consumption.

The on-farm use of stillage must be calculated on the basis of the number of animals that are normally kept and the quantity of stillage they can consume.

The potential for sale of stillage must be computed on the basis of letters of intent to purchase, not just on the existence of a local feedlot. The value of stillage will never exceed the directly corresponding cost of protein from other sources.

Direct on-farm use of carbon dioxide is limited; its principal value may come from sales. If Jerusalem artichokes, sorghum, or sugarcane are used, the bagasse and fiber that remain after the sugar is removed may be sufficient to supply the entire energy requirements of the ethanol plant. This value should be calculated in terms of the next less expensive source of fuel.

PRODUCTION POTENTIAL

Feedstocks

The mix of feedstocks determine in part the actual production potential. Chapter IV discusses the use and production of the various feedstocks individually and in

Stillage can be Used as a Protein Supplement When Mixed With the Proper Quantities of Grain and Forage

combination. The guidance offered in that chapter will help define the sizing of the plant from the viewpoint of output once the potential of the available feedstocks is determined. Additional feedstocks may also be purchased and combined with products available on-site.

Water

Significant amounts of water are used in the ethanol production process (about 16 gallons of water per gallon of ethanol produced). This demand includes requirements for generating steam, cooling, and preparing mashes. Also, it may be desirable to grow a crop not normally produced in the area. If additional irrigation water is necessary for this crop, the increment must be included, but it is likely that stillage liquids can be directly applied to fulfill this need.

Heat Sources

Heat is required in the conversion of feedstocks to ethanol, primarily in cooking, distillation, and stillage drying. An accurate assessment must be made to determine the type and quantity of available heat sources. Waste materials can contribute as energy sources and, from a national energy perspective, the use of petroleum fuels is not desirable. In some cases, other renewable sources of energy such as methane, solar, wind, and geothermal may be used as supplements.

EQUIPMENT SELECTION

The determination of the best equipment that can be obtained to fill the defined production needs is based on the operation's financial constraints and the labor and/or product compromises that can be made. All the options must be considered in relation to each other rather than independently.

The following variables related to equipment selection affect the decision to produce.

Labor Requirements

The availability of labor determines the schedule of plant operations and the degree of automation required. Labor availability is determined from normal farming routine and the disruptions which are tolerable.

Investment/Financing

Financing is a pivotal factor in the decision to produce. The options chosen depend initially upon capital and operating costs (which are in turn dependent upon plant size), and on individual financial situations. The potential income from the operation is the second line of consideration. Though no less important than the first, an inability to qualify for capital financing makes consideration of succeeding concerns a futile exercise.

Maintenance

Equipment maintenance varies in relation to the type of component and its use. In general, it can be assumed that the highest quality equipment will cost the most. Critical components should be identified and investments concentrated there. Noncritical or easily replaceable components can be less expensive. Routine maintenance should not interfere with production schedules.

Regulations

State and federal environmental protection standards must be observed. In addition, the Bureau of Alcohol, Tobacco, and Firearms (BATF) bonding requirements and regulations must be met. (See Appendix B for more information.)

Intended Use

Equipment must be selected that is capable of producing the quality, quantity, and form of coproducts dictated by the intended market.

Ethanol Feedstocks can be Stored in Conventional Facilities

Form of Coproducts

The form and amounts of coproducts will dictate the type and size of equipment. If stillage is to be sold in wet form, the only required equipment may be a storage tank. If stillage is to be dried, then screens, driers, and additional dry storage space will be necessary. If carbon dioxide is to be used or collected and sold, equipment for this will be needed.

Safety

Ethanol is extremely flammable and must be handled accordingly. Ignition sources must be isolated from all possible ethanol leaks. This isolation requirement affects either plant layout or equipment selection. The proper handling of acids and bases mandates particular types of construction materials.

Heat Sources

The type or types of fuels available to the operation will dictate the type of equipment necessary to convert this fuel into the required heat source.

Feedstock Mix

The desired feedstock mix will define the feed preparation equipment necessary (e.g., the production of ethanol from corn requires different front-end processing than sugar beets). Since it may be desirable to process more than one feedstock concurrently, additional equipment may be required in the processing step.

FINANCIAL REQUIREMENTS

Considerations to Proceed

Once the considerations for equipment selection are completed, the capital and operating costs may be roughly computed.

The capital cost considerations are:

- equipment,

- real estate and buildings,

- permits and licenses, and

- availability of financing.

The operating costs are:

- labor,

- cost of money,

- insurance,

- chemicals, enzymes, additives,

Agricultural Residue can be Collected in Large Round Bales for Storage

- fuel,

- feedstocks,

- costs of delivery, and

- bonds.

These considerations are then compared to the specific financial situation of the individual. If the results of this comparison are not acceptable, then other options in equipment specifications and plant size must be considered. If all possibilities result in an unfavorable position, the decision to produce is no. If a favorable set of conditions and specifications can be devised, detailed design considerations should be examined (see Chapter V, Plant Design) and an appropriate organization and financial plan developed (see Chapter VI, Business Plan).

DECISION AND PLANNING WORKSHEETS

The following questions are based on the considerations involved in deciding to proceed with development of a small-scale fermentation ethanol plant. Questions 1–28 are concerned with determining the potential market and production capability; questions 9, 20, and 29–47 examine plant size by comparing proposed income and savings with current earnings; questions 48–53 look at plant costs; questions 54–69 relate to financial and organizational requirements; and questions 71–85 examine financing options.

The final decision to produce ethanol is the result of examining all associated concerns at successively greater levels of detail. Initially a basic determination of feasibility must be made and its results are more a "decision to proceed with further investigation" than an ultimate choice to build a plant or not. This initial evaluation of feasibility is performed by examining: (1) the total market (including on-farm uses and benefits) for the ethanol and coproducts; (2) the actual production potential; (3) the approximate costs for building and operating a plant of the size that appropriately fits the potential market and the production potential; (4) the potential for revenues, savings, or indirect benefits; and (5) personal financial position with respect to the requirements for this plant. There are several points during the course of this evaluation that result in a negative answer. This does not necessarily mean that all approaches are infeasible. Retracing a few steps and adjusting conditions may establish favorable conditions; however, adjustments must be realistic, not overly optimistic. Similarly, completion of the exercise with a positive answer is no guarantee of success, it is merely a positive preliminary investigation. The real work begins with specifics.

Market Assessment

1. List equipment that runs on gasoline and estimate annual consumption for each.

	Equipment	Fuel Consumption	
a.	_____	_____	gal/yr
b.	_____	_____	gal/yr
c.	_____	_____	gal/yr
d.	_____	_____	gal/yr
e.	_____	_____	gal/yr
	TOTAL	_____	gal/yr

2. List the equipment from Question 1 that you intend to run on a 10%–EtOH/90%–gasoline blend.

	Equipment	Fuel Consumption	
a.	_____	_____	gal/yr
b.	_____	_____	gal/yr
c.	_____	_____	gal/yr
d.	_____	_____	gal/yr
e.	_____	_____	gal/yr
	TOTAL	_____	gal/yr

(Throughout these worksheets ethanol is abbreviated EtOH)

3. Take the total from Question 2 and multiply by 10% to obtain the quantity of EtOH to supply your own gasohol needs.

_____ × 0.1 = _____ gal EtOH/yr

4. List the equipment from Question 1 that you are willing to modify for straight EtOH fuel.

Equipment	Fuel Consumption	
a. _____	_____	gal/yr
b. _____	_____	gal/yr
c. _____	_____	gal/yr
d. _____	_____	gal/yr
e. _____	_____	gal/yr
TOTAL	_____	gal/yr

5. Take the total from Question 4 and multiply by 120% to obtain the quantity of EtOH for use as straight fuel in spark-ignition engines.

_____ gal/yr × 1.2 = _____ gal EtOH/yr

6. List your pieces of equipment that operate on diesel fuel.

Equipment	Diesel Fuel Consumption	
a. _____	_____	gal/yr
b. _____	_____	gal/yr
c. _____	_____	gal/yr
d. _____	_____	gal/yr
e. _____	_____	gal/yr
TOTAL	_____	gal/yr

7. List the equipment from Question 6 that you will convert to dual-injection system for 50% EtOH/50% diesel fuel blend.

Equipment	Diesel Fuel Consumption	
a. _____	_____	gal/yr
b. _____	_____	gal/yr
c. _____	_____	gal/yr
d. _____	_____	gal/yr
e. _____	_____	gal/yr
TOTAL	_____	gal/yr

8. Take the total from Question 7 and multiply by 0.5 to obtain the quantity of EtOH required for dual-injection system equipment.

_____ gal/yr × 0.5 = _____ gal EtOH/yr

9. Total the answers from Questions 3, 5, and 8 to determine your total annual on-farm EtOH consumption potential.

_____ gal EtOH/yr + _____ gal EtOH/yr + _____ gal EtOH/yr = _____ gal EtOH/yr

10. List the number of cattle you own that you intend to feed stillage.

a. _____ Feeder Calves

_____ Mature Cattle

A mature cow can consume the stillage from 1 gallon of ethanol production in 1 day. A feeder calf can consume the stillage from 0.7 gallon of ethanol production in 1 day. Multiply the number of feeder calves by 0.7. Add this product to the number of mature cattle to obtain the daily maximum EtOH production rate for which stillage can be consumed by cattle.

b. _____ Feeder Calves × 0.7 + _____ Mature Cattle = _____ gal/day

11. List the number of cattle that neighbors and/or neighboring feedlots own which they will commit to feed your stillage at full ration.

_____ Feeder Calves

_____ Mature Cattle

_____ Feeder Calves × 0.7 + _____ Mature Cattle = _____ gal/day

12. Total the answers from Questions 10 and 11 to determine the equivalent daily EtOH production rate for which the stillage can be consumed by cattle.

_____ gal/day + _____ gal/day = _____ gal/day

13. Determine the number of pigs you own that you can feed stillage.

a. _____ Pigs

Determine the number of pigs owned by neighbors or nearby pig feeders that can be committed to feeding your stillage at full ration.

b. _____ Neighbor Pigs

Total the results from a and b.

a + b. _____ Total Pigs

14. Multiply total from Question 13 by 0.4 to obtain equivalent daily EtOH production for which stillage can be consumed by pigs.

_____ Pigs × 0.4 = _____ gal/day

15. Repeat the exercise in Question 13 for sheep.

       a. _____ Sheep Owned

       b. _____ Neighbors Sheep

  a + b. _____ Total Sheep

16. Multiply total from Question 15 by the quantity of linseed meal normally fed every day to sheep in order to obtain the equivalent daily EtOH production rate for which stillage can be consumed by sheep.

_____ Sheep × _____ = gal/day

17. Repeat the excercise in Question 13 for poultry. Poultry can consume less than 0.05 lb of distillers' dried grains per day. This corresponds to about 0.07 gallons of whole stillage per day. Unless the poultry operation is very large, it is doubtful that this market can make any real contribution to consumption.

       a. _____ Poultry Owned

       b. _____ Neighbor's Poultry

  a + b. _____ Total Poultry

18. Take total from Question 17 and multiply by 0.05 to obtain the equivalent daily EtOH production rate for which stillage can be consumed by poultry.

_____ Poultry × 0.05 gal/day = _____ gal/day

19. Total the answers from Questions 11, 12, 14, 16, and 18 to obtain the total equivalent daily EtOH production rate for which stillage can be consumed by local livestock.

_____ gal/day + _____ gal/day + _____ gal/day + _____ gal/day + _____ gal/day = _____ gal/day

20. Multiply the total from Question 19 by 365 to obtain the total annual EtOH production for which the stillage will be consumed.

_____ gal/day × 365 = _____ gal/yr

Compare the answer from Question 20 to the answer from Question 9. If the answer from Question 20 is larger than the answer from Question 9, all of the stillage produced can be consumed by local livestock. This is the first production-limiting consideration. If the answer to Question 20 is smaller than the answer for Question 11, a choice must be made between limiting production to the number indicated by Question 20 or purchasing stillage processing equipment.

21. Survey the local EtOH purchase market to determine the quantity of EtOH that they will commit to purchase.

	High Proof		Anhydrous	
a. Dealers	_____	gal/yr	_____	gal/yr
b. Local Dist.	_____	gal/yr	_____	gal/yr
c. Regional Dist.	_____	gal/yr	_____	gal/yr
d. Other Farmers	_____	gal/yr	_____	gal/yr
e. Trans. Fleets	_____	gal/yr	_____	gal/yr
f. Fuel Blenders	_____	gal/yr	_____	gal/yr
TOTAL	_____	gal/yr	_____	gal/yr

22. Combine the answers from Questions 9 and 21 to determine annual market for EtOH.

_____ gal/yr + _____ gal/yr = _____ gal/yr

This is the ethanol market potential. It is not necessarily an appropriate plant size.

Production Potential

23. Which of the following potential EtOH feedstocks do you now grow?

		Annual		
	Acres	Yield/Acre	Production	
a. Corn	_____	_____	_____	bu/yr
b. Wheat	_____	_____	_____	bu/yr
c. Rye	_____	_____	_____	bu/yr
d. Barley	_____	_____	_____	bu/yr
e. Rice	_____	_____	_____	bu/yr
f. Potatoes	_____	_____	_____	cwt/yr
g. Sugar Beets	_____	_____	_____	tons/yr
h. Sugarcane	_____	_____	_____	tons/yr
i. Sweet Sorghum	_____	_____	_____	tons/yr

24. Do you have additional uncultivated land on which to plant more of any of these crops?

	Anticipated Acres	Potential Yield/Acre	Additional Annual Production	
a. Corn	_____	_____	_____	bu/yr
b. Wheat	_____	_____	_____	bu/yr
c. Rye	_____	_____	_____	bu/yr
d. Barley	_____	_____	_____	bu/yr
e. Rice	_____	_____	_____	bu/yr
f. Potatoes	_____	_____	_____	cwt/yr
g. Sugar Beets	_____	_____	_____	tons/yr
h. Sugarcane	_____	_____	_____	tons/yr
i. Sweet Sorghum	_____	_____	_____	tons/yr

25. Can you shift land from production of any of the crops not mentioned in Question 24 to increase production of one that is? If so, calculate the potential increase as in Question 24.

Crop	Anticipated Acres	Potential Yield/Acre	Additional Annual Production
_____	_____	_____	_____
_____	_____	_____	_____

26. Add the annual production values separately for each crop from Questions 22, 23, and 24. (This procedure can be used for other crops; however, reliable data for other crops are not available at this time.)

	Cereal Grains (combine totals) bu/yr	Potatoes cwt/yr	Sugar Beets ton/yr
a.	_____	_____	_____
b.	_____	_____	_____
c.	_____	_____	_____
TOTAL	_____	_____	_____
	Column I	Column II	Column III

27. Multiply the Question 26 answers from:

a. Column I by 2.5 to obtain annual potential EtOH production from cereal grains;

_____ bu/yr × 2.5 gal/bu = _____ gal EtOH/yr

b. Column II by 1.4 gal/cwt to obtain annual potential EtOH production from potatoes;

_____ cwt/yr × 1.4 gal/cwt = _____ gal EtOH/yr

c. Column III by 20 gal/ton to obtain annual potential EtOH production for sugar beets.

_____ ton/yr × 20 gal/ton = _____ gal EtOH/yr

28. Total the answers from Question 27a, 27b, and 27c to determine total *potential* production capability. (This is not necessarily the plant size to select, as the following series of questions demonstrates.)

_____ gal/yr + _____ gal/yr + _____ gal/yr = _____ gal/yr

If the answer to Question 28 is greater than the answer to Question 22, the *maximum* size of the plant would be the value from Question 22.

Plant Size

Neither the size of the market nor the production potential are sufficient to determine the appropriate plant size although they do provide an upper limit. A good starting point is to fill your own fuel needs (answer to Question 9) and not exceed local stillage consumption potential (answer to Question 20). Since the latter is usually larger and the equipment for treatment of stillage introduces a significant additional cost, the value from Question 20 is a good starting point. Now the approximate revenues and savings must be compared to current earnings from the proposed ethanol feedstock to determine if there is any gain in value by building an ethanol plant. Assume all feedstock costs are charged to production of EtOH.

Fuel Savings

29. Multiply the total of Questions 3 and 5 by the current price you pay for gasoline in $/gal.

(_____ gal/yr + _____ gal/yr) × _____ $/gal = _____ $/yr

This is the savings from replacing gasoline with EtOH.

30. Multiply the answer from Question 8 by the current price you pay for diesel fuel in $/gal.

_____ gal/yr × _____ $/yr = _____ $/yr

This is the savings from replacing diesel fuel with EtOH.

31. Total Questions 29 and 30 to obtain the fuel savings.

_____ $/yr + _____ $/yr = _____ $/yr

Feed Savings

32. Total the answers from Questions 10b, 14, 16, and 18.

_____ gal/day + _____ gal/day + _____ gal/day + _____ gal/day = _____ gal/day

33. Total the answers from Questions 11, 14, 16, and 18.

_____ gal/day + _____ gal/day + _____ gal/day + _____ gal/day = _____ gal/day

34. Multiply the answer to Question 32 by 6.8 to obtain the dry mass of high-protein material represented by the whole stillage fed (if using cereal grain feedstock).

_____ gal/day × 6.8 lb dry mass/gal EtOH = _____ lb dry mass/day

35. Multiply the answer to Question 34 by the protein fraction (e.g., 0.28 for corn) of the stillage on a dry basis.

_____ lb dry mass/day × _____ = _____ lb protein/day

(protein fraction)

36. a. Determine the cost (in $/lb protein) of the next less expensive protein supplement and multiply this number by the answer to Question 35 (answer this question only if you buy protein supplement).

_____ $/lb protein × _____ lb protein/day = _____ $/day

b. Multiply the answer to Question 36a by 365 (or the number of days per year you keep animals on protein supplement) to obtain annual savings in protein supplement.

_____ $/day × 365 days/yr = _____ $/yr

Production Savings

37. a. Determine the cost of production of high-protein feeds on your farm in $/lb dry mass and multiply by the protein fraction of each to obtain your actual cost of producing protein for feeding on-farm.

_____ $/lb dry mass × _____ = _____ $/lb protein

(protein fraction)

b. Multiply the answer to Question 37a by the answer to Question 35 (or by the amount of protein you actually produce on-farm: quantity in lbs times protein fraction, whichever is smaller) to obtain potential protein.

www.KnowledgePublications.com

_____ $/lb protein × _____ lb protein/day = _____ $/day

c. Multiply the answer to Question 37b by the number of days you keep animals on protein supplement during the year up to 365.

_____ $/day × _____ days/yr = _____ $/yr

38. Total the answers from Questions 36b and 37c.

_____ $/yr + _____ $/yr = _____ $/yr

This is the total animal feed savings you will realize each year.

Revenues

39. a. Multiply the answer from Question 28 by the reasonable market value of the stillage you produce.

_____ gal/day × _____ $/gal = _____ $/day

b. Multiply the answer obtained in Question 39a by the number of days during the year that this quantity of stillage can be marketed, up to 365.

_____ $/day × _____ days/yr = _____ $/yr

This is the total stillage sales you will realize each year.

40. Total the answers from Questions 38 and 39b.

_____ $/yr + _____ $/yr = _____ $/yr

This is the total market value of the stillage you will produce.

41. Subtract the answer to Question 9 from the answer to Question 20 to obtain the EtOH production potential that remains for sale.

_____ gal/yr − _____ gal/yr = _____ gal/yr

42. Multiply the answer from Question 41 by the current market value for ethanol.

_____ gal/yr × _____ $/gal = _____ $/yr

This is the annual ethanol sales potential.

43. Total the answers from Questions 31, 38, 40, and 42 to obtain the total revenues and savings from this production rate.

_____ $/yr + _____ $/yr + _____ $/yr + _____ $/yr = _____ $/yr

44. Divide the answer to Question 20 by:

a. 2.5 gal/bu if the feedstock to be used is cereal grain.

_____ gal/yr ÷ 2.5 gal/bu = _____ bu/yr

b. 1.4 gal/cwt if the feedstock to be used is potatoes.

_____ gal/yr ÷ 1.4 gal/cwt = _____ cwt/yr

c. 20 gal/ton if the feedstock to be used is sugar beets.

_____ gal/yr ÷ 20 gal/ton = _____ tons/yr

45. Multiply:

a. The answer from Question 44a by the appropriate market value for cereal grains to obtain the potential earnings for direct marketing with EtOH production;

_____ bu/yr × _____ $/bu = _____ $/yr

b. The answer from Question 44b by the appropriate market value for potatoes;

_____ cwt/yr × _____ $/cwt = _____ $/yr

c. The answer from Question 44c by the appropriate market value for sugar beets.

_____ tons/yr × _____ $/ton = _____ $/yr

46. Total the answers from Questions 45a, 45b, and 45c to obtain the potential earnings from directly marketing crops without making EtOH.

_____ $/yr + _____ $/yr + _____ $/yr = _____ $/yr

Compare the answers from Questions 46 and 43. If Question 46 is as large, or nearly as large as the answer from Question 43, the construction of an ethanol plant of this size cannot be justified on a purely economic basis. Consider scaling down to a size that fills your own fuel needs and recompute Questions 29 through 46. If Question 43 is considerably larger (2 to 3 times) than Question 46, you can consider increasing your plant size within the bounds of the answers to Question 22 (market) and Question 28 (production potential). Care must be taken to assess local competition and market share as you expand plant size.

If a market share exists or if there is good reason to believe that you can acquire a share by superior techniques, the initial plant sizing must accurately reflect this realistic market share.

47. a. Multiply the initial plant production capacity (in gallons EtOH/hr) by 16 gallons of water per gallon EtOH production capacity.

_____ gal EtOH/hr × 16 gal H₂O/gal EtOH = _____ gal H₂O/hr.

b. Can the answer to Question 47 be realistically achieved in your area? If yes, no adjustment to chosen plant size needs to be made to account for water availability. If no, reduce plant size to realistically reflect available water.

Approximate Costs of Plant

The cost of the equipment you choose will be a function of the labor available, the maintenance required, the heat source selected, and the type of operating mode.

Labor Requirements

How much time during the normal farming routine can you dedicate to running the ethanol plant?

48. a. Do you have any hired help or other adult family members, and if so, how much time can he/she dedicate to running the ethanol plant?

    b. Can you or your family or help dedicate time at periodic intervals to operating the ethanol plant?

If labor is limited, a high degree of automatic control is indicated.

Maintenance

49. What are your maintenance capabilities and equipment?

Heat Source

Determine the least expensive heat source available.

50. Select a plant design that accomplishes your determined production rate and fits your production schedule.

51. List all of the plant components and their costs

    a) storage bins $ _____

    b) grinding mill $ _____

    c) meal hopper $ _____

    d) cookers $ _____

    e) fermenters $ _____

    f) distillation columns $ _____

    g) storage tanks (product and coproduct) $ _____

    h) pumps $ _____

    i) controllers $ _____

    j) pipes and valves $ _____

    k) metering controls $ _____

    l) microprocessors $ _____

    m) safety valves $ _____

n) heat exchangers $ _____

o) instrumentation $ _____

p) insulation $ _____

q) boiler $ _____

r) fuel handling equipment $ _____

s) feedstocks handling equipment $ _____

t) storage tanks (stillage) $ _____

u) stillage treatment equipment
    (screen, dryers, etc.) $ _____

v) CO_2 handling equipment $ _____

w) ethanol dehydration equipment $ _____

TOTAL $ _____

52. Determine operating requirements for cost.

Plant capacity = _____ gallons of anhydrous ethanol per hour.

Production = _____ gallons per hour × hours of operation per year = _____ gal/yr.

Feed materials = Production _____ gal/yr ÷ _____ gal/bu = _____ bu/yr.

	$/yr	$/gal

a. Operating Costs
Feed materials
  Grain ($/bu ÷ gal
  anhydrous ethanol/bu
  = $/gal.)
	_____	_____

or ($/bu × bu/yr = $/yr.)	_____	_____
Supplies Enzymes	_____	_____
Other	_____	_____
Fuel for plant operation	_____	_____
Waste disposal	_____	_____
Operating labor (operating crew × hrs of operation per year × $/hr = $/year)	_____	_____
Total Operating Costs	_____	_____

b. Maintenance Costs
    Routine scheduled maintenance _____ _____

    Labor (Maintenance crew staff
    × hrs/yr × $/hr) _____ _____

    Supplies and replacement parts _____ _____

    Maintenance equipment rental _____ _____

Unscheduled Maintenance (Estimated) _____ _____

    Labor _____ _____

    Supplies _____ _____

    Maintenance equipment _____ _____

Total Maintenance Costs ============== ==============

c. Capital or Investment Costs
    Plant equipment costs _____ _____

    Land _____ _____

    Inventory
        Grain _____ _____

        Supplies _____ _____

        Ethanol _____ _____

        Spare parts _____ _____

    Total ============== ==============

    Taxes _____ _____

    Insurance _____ _____

    Depreciation _____ _____

    Interest on loan or mortgage _____ _____

    Total Capital or Investment Costs ============== ==============

    TOTAL COSTS (Totals of a, b and c) ============== ==============

Financial Requirements

53. Capital Costs

Item	Cost Estimate	Considerations
Real estate	_____	_____

Item	Cost Estimate	Considerations
Buildings	_____	_____
Equipment	_____	_____
Business formation	_____	_____
Equipment installation	_____	_____
Licensing costs	_____	_____

54. Operating Costs

Item	Cost Estimate	Considerations
Labor	_____	_____
Maintenance	_____	_____
Taxes	_____	_____
Supplies	_____	Includes raw materials, additives, enzymes, yeast, and water.
Delivery	_____	_____
Expenses	_____	Includes electricity and fuel(s).
Insurance	_____	_____
Interest on debt	_____	Includes interest on long- and short-term loans.
Bonding	_____	_____

55. Start-Up Working Capital

Item	Cost Estimate	Considerations
Mortgage	_____	Principle payments only, for first few months.
Cash to carry accounts receivable for 60 days	_____	_____
Cash to carry a finished goods inventory for 30 days	_____	_____
Cash to carry a raw material inventory for 30 days	_____	_____

56. Working Capital

Item	Cost Estimate	Considerations
Mortgage	_____	Principle payments only.

Assets (Total Net Worth)

57. List all items owned by the business entity operating the ethanol plant.

Item	Value
_____	_____
_____	_____
_____	_____
_____	_____
_____	_____
_____	_____
_____	_____
_____	_____
_____	_____
_____	_____
_____	_____
_____	_____
_____	_____

Organizational Form

58. Are you willing to assume the costs and risks of running your own EtOH production facility? _____

59. Are you capable of handling the additional taxes and debts for which you will be personally liable as a single proprietor? _____
Includes interest on long- and short-term loans.

60. Is your farm operation large enough or are your potential markets solid enough to handle an EtOH production facility as a single proprietor? _____

61. Is your credit alone sufficient to provide grounds for capitalizing a single proprietorship? _____

62. Will a partner(s) enhance your financial position or supply needed additional skills? _____

63. a. Do you need a partner to get enough feedstock for your EtOH production facility? _____

    b. Are you willing to assume liabilities for product and partner? _____

64. Is your intended production going to be of such a scale as to far exceed the needs for your own farm or several neighboring farms? _____

65. Do you need to incorporate in order to obtain adequate funding? _____

66. Will incorporation reduce your personal tax burden? _____

67. Do you wish to assume product liability personally? _____

68. How many farmers in your area would want to join a cooperative? _____

69. Do you plan to operate in a centralized location to produce EtOH for all the members? _____

70. Is your main reason for producing EtOH to service the needs of the cooperative members, others, or to realize a significant profit? _____

Financing

If you are considering borrowing money, you should have a clear idea of what your chances will be beforehand. The following questions will tell you whether debt financing is a feasible approach to your funding problem.

71. a. How much money do you already owe? _____

    b. What are your monthly payments? _____

72. How much capital will you have to come up with yourself in order to secure a loan? _____

73. Have you recently been refused credit? _____

74. a. How high are the interest rates going to be? _____

    b. Can you cover them with your projected cash flow? _____

75. If the loan must be secured or collateralized, do you have sufficient assets to cover your debt? _____

If you are already carrying a heavy debt load and/or your credit rating is low, your chances of obtaining additional debt financing is low and perhaps you should consider some other type of financing. Insufficient collateral, exorbitant interest rates, and low projected cash flow are also negative indicators for debt financing.

The choice between debt and equity financing will be one of the most important decisions you will have to face since it will affect how much control you will ultimately have over your operation. The following questions deal with this issue, as well as the comparative cost of the two major types of financing.

76. How much equity do you already have? _____

77. Do you want to maintain complete ownership and control of your enterprise? _____

78. Are you willing to share ownership and/or control if it does not entail more than a minority share? _____

79. Will the cost of selling the stock (broker's fee, bookkeeper, etc.) be more than the interest you would have to pay on a loan? _____

If you are reluctant to relinquish any control over your operation, you would probably be better off seeking a loan. On the other hand, if your chances of obtaining a loan are slim, you might have to trade off some personal equity in return for a better borrowing position.

80. Do you have other funds or materials to match with federal funds? (It is usually helpful.) _____

81. Do you live in a geographical area that qualifies for special funds? _____
82. Will you need continued federal support at the end of your grant period? _____

83. Are you going to apply for grant funds as an individual, as a nonprofit corporation, or as a profit corporation? _____

84. Are you a private nonprofit corporation? _____

85. Is there something special about your alcohol facility that would make it attractive to certain foundations? _____

You should now have a good idea as to where you are going to seek your initial funding. Remember that most new businesses start up with a combination of funding sources. It is important to maintain a balance that will give you not only sufficient funding when you need it, but also the amount of control over your operation that you would like to have.

Completion of these worksheets can lead to an initial decision on the feasibility to proceed. However, this should not be construed as a final decision, but rather a step in that process.

If the financial requirements are greater than the capability to obtain financing, it does not necessarily mean the entire concept will not work. Rather, the organizational form can be reexamined and/or the production base expanded in order to increase financing capability.

REFERENCES

1. Reed, Thomas. "Alcohol Fuels." Special Hearing of the U.S. Senate Committee on Appropriations; Washington, D.C.: January 31, 1978; pp. 194-205. U.S. Government Printing Office. Stock Number 052-070-04679-1.

2. Jawetz, Pincas. "Improving Octane Values of Unleaded Gasoline Via Gasohol." *Proceedings of the 14th Intersociety Energy Conversion Engineering Conference.* Volume I, pp. 301-302; abstract Volume II, p. 102; Boston, MA: August 5-10, 1979. Available from the American Chemical Society, 1155 Sixteenth Street NW, Washington, D.C. 20036.

3. Panchapakesan, M.R.; et al. "Factors That Improve the Performance of an Ethanol Diesel Oil Dual-Fuel Engine." International Symposium on Alcohol Fuel Technology–Methanol and Ethanol. Wolfsburg, Germany; November 21-23, 1977. CONF-771175.

4. *Ethanol Chemical and Engineering News.* October 29, 1979; p. 12.

5. Sharma, K.D. "Present Status of Alcohol and Alcohol-Based Chemical Industry in India." Workshop on Fermentation Alcohol for Use as Fuel and Chemical Feedstock in Developing Countries. United Nations International Development Organization, Vienna, Austria; March 1979. Paper no. ID/WG.293/14 UNIDO.

6. Ribeiro, Filho F. A. "The Ethanol-Based Chemical Industry in Brazil." Workshop on Fermentation Alcohol for Use as Fuel and Chemical Feedstock in Developing Countries. United Nations Industrial Development Organization, Vienna, Austria; March 1979. Paper no. ID/WG.293/4 UNIDO.

CHAPTER III
Basic Ethanol Production

CHAPTER III
Basic Ethanol Production

The production of ethanol is an established process. It involves some of the knowledge and skill used in normal farm operations, especially the cultivation of plants. It is also a mix of technologies which includes microbiology, chemistry, and engineering. Basically, fermentation is a process in which microorganisms such as yeasts convert simple sugars to ethanol and carbon dioxide. Some plants directly yield simple sugars; others produce starch or cellulose that must be converted to sugar. The sugar obtained must be fermented and the beer produced must then be distilled to obtain fuel-grade ethanol. Each step is discussed individually. A basic flow diagram of ethanol production is shown in Figure III-1.

PREPARATION OF FEEDSTOCKS

Feedstocks can be selected from among many plants that either produce simple sugars directly or produce starch and cellulose. The broad category of plants which this includes means there is considerable diversity in the initial processing, but some features are universal:

- simple sugars must be extracted from the plants that directly produce them;

- starch and cellulose must be reduced from their complex form to basic glucose; and

- stones and metallic particles must be removed.

The last feature must be taken care of first. Destoning equipment and magnetic separators can be used to remove stones and metallic particles. Root crops require other approaches since mechanical harvesters don't differentiate between rocks and potatoes or beets of the same size. Water jets or flumes may be needed to accomplish this.

The simple sugars from such plants as sugarcane, sugar beets, or sorghum can be obtained by crushing or pressing the material. The low sugar bagasse and pulp which remain after pressing can be leached with water to remove residual sugars. The fibrous cellulosic material theoretically could be treated chemically or enzymatically to produce more sugar. However, no commercially available process currently exists.

Commonly used starchy feedstocks are grains and potatoes. Starch is roughly 20% amylose (a water-soluble carbohydrate) and 80% amylopectin (which is not soluble in water). These molecules are linked together by means of a bond that can be broken with relative ease. Cellulose, which is also made up of glucose, differs from starch mainly in the bond between glucose units.

Starch must be broken down because yeast can only act on simple sugars to produce ethanol. This process requires that the material be broken mechanically into the smallest practical size by milling or grinding, thereby breaking the starch walls to make all of the material available to the water. From this mixture, a slurry can be prepared and it can be heated to temperatures high enough to break the cell walls of the starch. This produces complex sugars which can be further reduced by enzymes to the desired sugar product.

Conversion of Starches by Enzymatic Hydrolysis

Consider the preparation of starch from grain as an example of enzymatic hydrolysis. The intent is to produce a 14% to 20% sugar solution with water and whole grain. Grain is a good source of carbohydrate, but to gain access to the carbohydrate, the grain must be ground. A rule of thumb is to operate grinders so that the resulting meal can pass a 20-mesh screen. This assures that the carbohydrate is accessible and the solids can be removed with a finer screen if desired. If the grain is not ground finely enough, the resultant lumpy material is not readily accessible for enzymatic conversion to sugar. The next step is to prepare a slurry by mixing the meal directly with water. Stirring should be adequate to prevent the formation of lumps and enhance enzyme contact with the starch (thus speeding liquefaction).

High-temperature and high-pressure processes may require a full time operator, thus making it difficult to integrate into farming operations. Therefore, when deciding which enzyme to purchase, consideration should be given to selecting one that is active at moderate temperature, i.e., 200° F (93° C), near-ambient pressure, and nearly neutral pH. The acidity of the slurry can be adjusted by addition of dilute basic solution (e.g., sodium hydroxide) if the pH is too low and addition of concentrated sulfuric acid or lactic acid if the pH is too high.

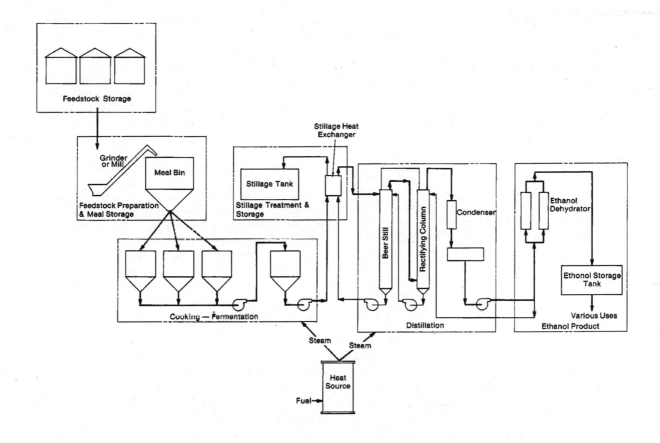

Figure III-1. Ethanol Production Flow Diagram

The enzyme should be added to the slurry in the proper proportion to the quantity of starch to be converted. If not, liquefaction ends up incomplete or takes too long to complete for practical operations. Enzymes vary in activity but thermophyllic bacterial amylases, which are commercially available, can be added at rates slightly greater than ¾ ounce per bushel of meal. Rapid dispersion of the dry enzyme is best accomplished by mixing a premeasured quantity with a small volume of warm water prior to addition to the slurry. Liquefaction should be conducted in the specific temperature range and pH suggested by the supplier of the specific enzyme used.

After the enzyme is added, the grain mash is heated to break the cell walls of the starch. However, the enzyme must be added before the temperature is raised because once the cell walls rupture, a gel forms and it becomes almost impossible to accomplish good mixing of the enzyme with the starch. The rupture of cell walls, which is caused by heating in hot water, is called gelatinization because the slurry (which is a suspension of basically insoluble material in water) is converted to a high-viscosity solution. Under slow cooking conditions and normal atmospheric pressure, gelatinization can be expected to occur around 140° F (60° C).

The temperature is then raised to the optimal functional range for the enzyme and held for a period of time sufficient to completely convert the starch to soluble dextrins (polymeric sugars). There are commercially available enzymes that are most active around 200° F (93° C) and require a hold time of 2½ hours if the proper proportion of enzyme is used. When this step is complete, the slurry has been converted to an aqueous solution of dextrins. Care must be taken to assure that the starch conversion step is complete because the conditions for the glucose-producing enzyme (glucoamylase), which is introduced in the next step, are significantly different from those for liquefaction.

The next step, saccharification, is the conversion of dextrins to simple sugars, i.e., glucose. The mash temperature is dropped to the active range of the glucoamylase, the enzyme used for saccharification, and the pH of the solution is adjusted to optimize conversion activity. The pH is a critical factor because the enzymatic activity virtually ceases when the pH is above 6.5. Glucoamylase is added in the proportion required to convert the amount of sugar available. Again, depending upon the variety selected and its activity, the actual required quantity of enzyme varies.

After the enzyme is added, the temperature of the mash must neither exceed 140° F (60° C) nor drop below 122° F (50° C) during the saccharification step or the enzyme activity is greatly reduced. The mash, as in the prior step, must be stirred continuously to assure intimate contact of enzyme and dextrin. The mash should be held at the proper temperature and pH until conversion of the dextrin to glucose is complete.

FERMENTATION

Fermentation is the conversion of an organic material from one chemical form to another using enzymes produced by living microorganisms. In general, these bacteria are classified according to their tolerance of oxygen. Those that use oxygen are called aerobic and those that do not are called anaerobic. Those that start with oxygen but continue to thrive after all of the available oxygen is consumed are called facultative organisms. The yeast used to produce ethanol is an example of this type of facultative anaerobe. The breakdown of glucose to ethanol involves a complex sequence of chemical reactions which can be summarized as:

$$C_6H_{12}O_6 \longrightarrow 2C_2H_5OH + 2CO_2 + heat$$

(glucose)   (ethanol)   (carbon dioxide)

Actual yields of ethanol generally fall short of predicted theoretical yields because about 5% of the sugar is used by the yeast to produce new cells and minor products such as glycerols, acetic acid, lactic acid, and fusel oils.

Yeasts are the microorganisms responsible for producing the enzymes which convert sugar to ethanol. Yeasts are single-cell fungi widely distributed in nature, commonly found in wood, dirt, plant matter, and on the surface of fruits and flowers. They are spread by wind and insects. Yeasts used in ethanol productions are members of the genus *Saccharomyces*. These yeasts are sensitive to a wide variety of variables that potentially affect ethanol production. However, pH and temperature are the most influential of these variables. *Saccharomyces* are most effective in pH ranges between 3.0 and 5.0 and temperatures between 80° F (27° C) and 90° F (35° C). The length of time required to convert a mash to ethanol is dependent on the number of yeast cells per quantity of sugar. The greater the number initially added, the faster the job is complete. However, there is a point of diminishing returns.

Yeast strains, nutritional requirements, sugar concentration, temperature, infections, and pH influence yeast efficiency. They are described as follows:

Yeast Strains

Yeasts are divided informally into top and bottom yeasts according to the location in the mash in which most of the fermentation takes place. The top yeasts, *Saccharomyces cerevisiae,* produce carbon dioxide and ethanol vigorously and tend to cluster on the surface of the mash. Producers of distilled spirits generally use top yeasts of high activity to maximize ethanol yield in the shortest time; producers of beer tend to use bottom yeasts which produce lower ethanol yields and require longer times to complete fermentation. Under normal brewing conditions, top yeasts tend to flocculate (aggregate together into clusters) and to separate out from the solution when fermentation is complete. The various strains of yeast differ considerably in their tendency to flocculate. Those strains with an excessive tendency toward premature flocculation tend to cut short fermentation and thus reduce ethanol yield. This phenomenon, however, is not singularly a trait of the yeast. Fermentation conditions can be an influencing factor. The cause of premature flocculation seems to be a function of the pH of the mash and the number of free calcium ions in solution. Hydrated lime, which is sometimes used to adjust pH, contains calcium and may be a contributory factor.

Nutritional Requirements

Yeasts are plants, despite the fact that they contain no chlorophyll. As such, their nutritional requirements must be met or they cannot produce ethanol as fast as desired. Like the other living things that a farmer cultivates and nurtures, an energy source such as carbohydrate must be provided for metabolism. Amino acids must be provided in the proper proportion and major chemical elements such as carbon, nitrogen, phosphorus, and others must be available to promote cell growth. Some species flourish without vitamin supplements, but in most cases cell growth is enhanced when B-vitamins are available. Carbon is provided by the many carbonaceous substances in the mash.

The nitrogen requirement varies somewhat with the strain of yeast used. In general, it should be supplied in the form of ammonia, ammonium salts, amino acids, peptides, or urea. Care should be taken to sterilize farm sources of urea to prevent contamination of the mash with undesired microbial strains. Since only a few species of yeasts can assimilate nitrogen from nitrates, this is not a recommended source of nitrogen. Ammonia is usually the preferred nitrogen form, but in its absence, the yeast can break up amino acids to obtain it. The separation of solids from the solution prior to fermentation removes the bulk of the protein and, hence, the amine source would be removed also. If this option is exercised, an ammonia supplement must be provided or yeast populations will not propagate at the desired rates and fermentation will take an excessive amount of time to complete. However, excessive amounts of ammonia in solution must be avoided because it can be lethal to the yeast.

Although the exact mineral requirements of yeasts cannot be specified because of their short-term evolutionary capability, phosphorus and potassium can be identified as elements of prime importance. Care should be taken not to introduce excessive trace minerals, because those which the yeast cannot use increase the osmotic pressure in the system. (Osmotic pressure is due to the physical imbalance in concentration of chemicals on either side of a membrane. Since yeasts are cellular organisms, they are enclosed by a cell wall. An excessively high osmotic pressure can cause the rupture of the cell wall which in turn kills the yeast.)

Sugar Concentration

There are two basic concerns that govern the sugar concentration of the mash: (1) excessively high sugar concentrations can inhibit the growth of yeast cells in the initial stages of fermentation, and (2) high ethanol concentrations are lethal to yeast. If the concentration of ethanol in the solution reaches levels high enough to kill yeast before all the sugar is consumed, the quantity of sugar that remains is wasted. The latter concern is the governing control. Yeast growth problems can be overcome by using large inoculations to start fermentation. *Saccharomyces* strains can utilize effectively all of the sugar in solutions that are 16% to 22% sugar while producing a beer that ranges from 8% to 12% ethanol by volume.

Temperature

Fermentation is strongly influenced by temperature, because the yeast performs best in a specific temperature range. The rate of fermentation increases with temperature in the temperature range between 80° F (27° C) and 95° F (35° C). Above 95° F (35° C), the rate of fermentation gradually drops off, and ceases altogether at temperatures above 109° F (43° C). The actual temperature effects vary with different yeast strains and typical operating conditions are generally closer to 80° F (27° C) than 95° F (35° C). This choice is usually made to reduce ethanol losses by evaporation from the beer. For every 9° F (5° C) increase in temperature, the ethanol evaporation rate increases 1.5 times. Since scrubbing equipment is required to recover the ethanol lost by evaporation and the cost justification is minimal on a small scale, the lower fermentation temperature offers advantages of simplicity.

The fermentation reaction gives off energy as it proceeds (about 500 Btu per pound of ethanol produced). There will be a normal heat loss from the fermentation tank as long as the temperature outside the tank is less than that inside. Depending upon the location of the plant, this will depend on how much colder the outside air is than the inside air and upon the design of the fermenter. In general, this temperature difference will not be sufficient to take away as much heat as is generated by the reaction except during the colder times

of the year. Thus, the fermenters must be equipped with active cooling systems, such as cooling coils and external jackets, to circulate air or water for convective cooling.

Infections

Unwanted microbial contaminants can be a major cause of reductions in ethanol yield. Contaminants consume sugar that would otherwise be available for ethanol production and produce enzymes that modify fermentation conditions, thus yielding a drastically different set of products. Although infection must be high before appreciable quantities of sugar are consumed, the rate at which many bacteria multiply exceeds yeast propagation. Therefore, even low initial levels of infection can greatly impair fermentation. In a sense, the start of fermentation is a race among the microorganisms present to see who can consume the most. The objective is the selective culture of a preferred organism. This means providing the conditions that are most favorable to the desired microorganism. As mentioned previously, high initial sugar concentrations inhibit propagation of *Saccharomyces cerevisiae* because it is not an osmophylic yeast (i.e., it cannot stand the high osmotic pressure caused by the high concentration of sugar in the solution). This immediately gives an advantage to any osmophylic bacteria present.

Unwanted microbes can be controlled by using commercially available antiseptics. These antiseptics are the same as those used to control infections in humans, but are less expensive because they are manufactured for industrial use.

DISTILLATION

The purpose of the distillation process is to separate ethanol from the ethanol-water mixture. There are many means of separating liquids comprised of two or more components in solution. In general, for solutions comprised of components of significantly different boiling temperatures, distillation has proved to be the most easily operated and thermally efficient separation technique.

At atmospheric pressure, water boils around 212° F (100° C) and ethanol boils around 172° F (77.7° C). It is this difference in boiling temperature that allows for distillative separation of ethanol-water mixtures. If a pan of an ethanol and water solution is heated on the stove, more ethanol molecules leave the pan than water molecules. If the vapor leaving the pan is caught and condensed, the concentration of ethanol in the condensed liquid will be higher than in the original solution, and the solution remaining in the pan will be lower in ethanol concentration. If the condensate from this step is again heated and the vapors condensed, the con-

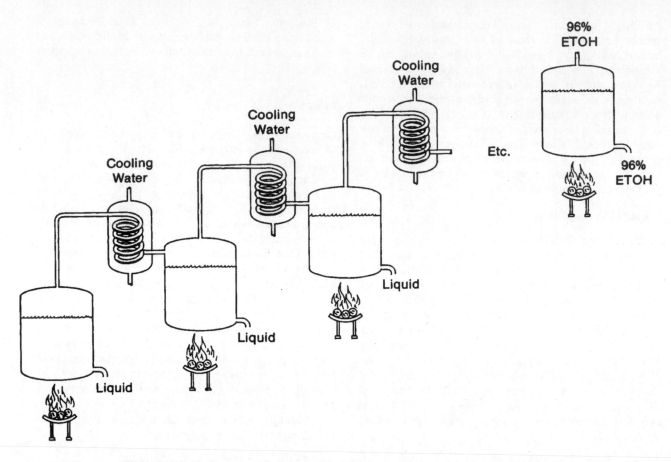

Figure III-2. Basic Process of Successive Distillation to Increase Concentration of Ethanol

centration of ethanol in the condensate will again be higher. This process could be repeated until most of the ethanol was concentrated in one phase. Unfortunately, a constant boiling mixture (azeotrope) forms at about 96% ethanol. This means that when a pan containing a 96%-ethanol solution is heated, the ratio of ethanol molecules to water molecules in the condensate remains constant. Therefore, no concentration enhancement is achieved beyond this point by the distillation method.

The system shown in Figure III-2 is capable of producing 96%-pure ethanol, but the amount of final product will be quite small. At the same time there will be a large number of products of intermediate ethanol-water compositions that have not been brought to the required product purity. If, instead of discarding all the intermediate concentrations of ethanol and water, they were recycled to a point in the system where the concentration was the same, we could retain all the ethanol in the system. Then, if all of these steps were incorporated into one vessel, the result would be a distillation column. The advantages of this system are that no intermediate product is discarded and only one external heating and one external cooling device are required. Condensation at one stage is affected when vapors contact a cooler stage above it, and evaporation is affected when liquid contacts a heating stage below.

Heat for the system is provided at the bottom of the distillation column; cooling is provided by a condenser at the top where the condensed product is returned in a process called reflux. It is important to note that without this reflux the system would return to a composition similar to the mixture in the first pan that was heated on the stove.

The example distillation sieve tray column given in Figure III-3 is the most common single-vessel device for carrying out distillation. The liquid flows down the tower under the force of gravity while the vapor flows upward under the force of a slight pressure drop.

The portion of the column above the feed is called the rectifying or enrichment section. The upper section serves primarily to remove the component with the lower vapor pressure (water) from the upflowing vapor, thereby enriching the ethanol concentration. The portion of the column below the feed, called the stripping section, serves primarily to remove or strip the ethanol from the down-flowing liquid.

Figure III-4 is an enlarged illustration of a sieve tray. In order to achieve good mixing between phases and to provide the necessary disengagement of vapor and liquid between stages, the liquid is retained on each plate

www.KnowledgePublications.com

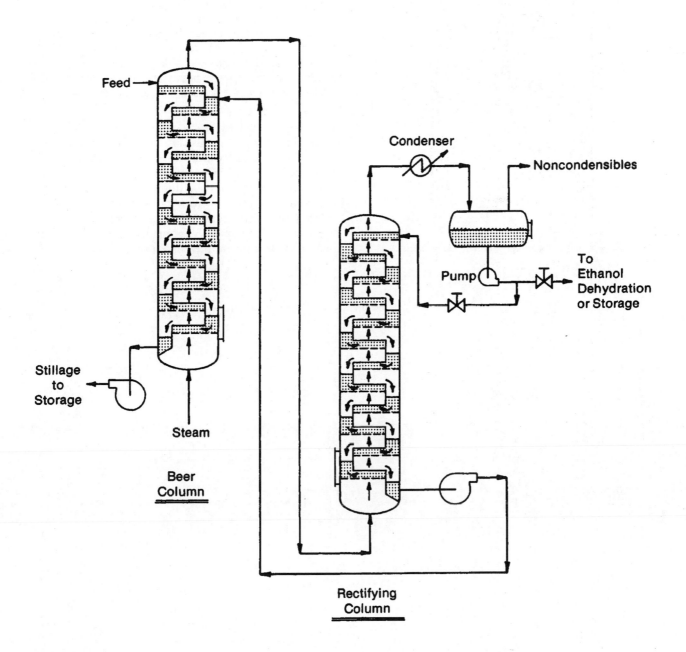

Feed

Condenser

Noncondensibles

Pump

To
Ethanol
Dehydration
or Storage

Stillage
to
Storage

Steam

Beer
Column

Rectifying
Column

Figure III-3. Schematic Diagram of Sieve Tray Distillation of Ethanol

Figure III-4. Enlarged Illustration of Sieve Tray

by a weir (a dam that regulates flow) over which the solution flows. The effluent liquid then flows down the downcomer to the next stage. The downcomer provides sufficient volume and residence time to allow the vapor-liquid separation.

It is possible to use several devices other than sieve tray columns to achieve the counter-current flow required for ethanol-water distillation. A packed column is frequently used to effect the necessary vapor-liquid contacting. The packed column is filled with solid material shaped to provide a large surface area for contact. Counter-current liquid and vapor flows proceed in the same way described for the sieve tray column.

Production of fuel-grade ethanol is a practical operation to include in farm activities. Texts in microbiology and organic chemistry portray it as a complex procedure, but this is not necessarily true. Fermentation is affected by a variety of conditions. The more care used in producing optimum conditions, the greater the ethanol yield. Distillation can range from the simple to the complex. Fortunately, the middle line works quite satisfactorily for on-farm ethanol production.

CHAPTER IV
Feedstocks

CHAPTER IV
Feedstocks

The previous chapter discussed the basic process for fermenting sugar into ethanol. The purposes of this chapter are (1) to describe the types of agricultural crops and crop residues that make up the feedstocks used in the production of ethanol; (2) to provide data on the yield of the three principal coproducts derived from fermentation of these feedstocks; and (3) to present agronomic and feedstock considerations of ethanol production.

TYPES OF FEEDSTOCKS

Biological production of ethanol is accomplished by yeast through fermentation of six-carbon sugar units (principally glucose). All agricultural crops and crop residues contain six-carbon sugars, or compounds of these sugars, and therefore can be used in the production of ethanol. Three different arrangements of the basic sugar units are possible, as seen in the three different types of agricultural feedstocks available for fermentation: sugar crops, starch crops, and lignocellulosic residues. The starch crops and lignocellulosic residues contain six-carbon sugar compounds which must be broken down into simple six-carbon sugar units before fermentation can take place.

Sugar Crops

In sugar crops, the majority of the six-carbon sugar units occur individually or in bonded pairs. Once a sugar crop has been crushed to remove the sugar, no additional processing is needed prior to fermentation since the six-carbon sugar units are already in a form that the yeast can use. This fact is both an advantage and a disadvantage. Preparation of the feedstock for fermentation involves comparatively low equipment, labor, and energy costs, since the only major steps involved are milling and extraction of the sugar. However, sugar crops tend to spoil easily. Numerous types of microorganisms (including the type of yeast that produces ethanol) thrive on these crops during storage because of their high moisture and sugar content. Therefore, steps must be taken during storage to slow the loss of sugar. The only proven storage method is evaporation of water from the sugar solution—an effective, but costly method in terms of equipment (evaporators) and energy. Sterilization of the juice by use of heat, chemicals, or ultrafiltration to remove microbes is currently under investigation [1].

The two sugar crops that have been cultivated in the United States for many years at a commercial level of production are sugarcane and sugar beets. Other alternative sugar crops that can be cultivated in the United States include sweet sorghum, Jerusalem artichokes, fodder beets, and fruits.

Sugarcane. Sugarcane is considered a favorable feedstock because of its high yield of sugar per acre (as high as 50 tons per acre per year) and a correspondingly high yield of crop residue, known as bagasse, that can be used as a fuel for production of process heat. The major drawback with this feedstock is the limited availability of land suitable for economical cultivation. Presently, only four states (Florida, Louisiana, Texas, and Hawaii) cultivate sugarcane.

Sugar Beets. Sugar beets are a much more versatile crop than sugarcane. They are presently grown in 19 states, and the potential for cultivation in other parts of the country is high because sugar beets tolerate a wide range of climatic and soil conditions. An important advantage of sugar beets is the comparatively high yield of crop coproducts: beet pulp and beet tops. Beet pulp, the por-

Sugar Beets are a Good Ethanol Feedstock

Sweet Sorghum Yields Grain and Sugar, Both of Which can be Used as Ethanol Feedstocks

tion of the root that remains after the sugar has been removed, is bulky and palatable and may be fed in either wet or dry form. Beet tops have alternative uses that include leaving them on the field for organic material (fertilizer) and as cover to lessen soil erosion.

Widespread expansion of sugar beet cultivation is limited to some extent by the necessity to rotate with nonroot crops, in order to lower losses caused by a buildup of nematodes, a parasitic worm that attacks root systems. A general guideline of one beet crop per 4-year period should be followed. None of the sugar beet crop coproducts are suitable for use as a boiler fuel.

Interest in ethanol production from agricultural crops has prompted research on the development of sugar crops that have not been cultivated on a widespread commercial basis in this country. Three of the principal crops now under investigation are sweet sorghum, Jerusalem artichokes, and fodder beets.

Sweet Sorghum. Sweet sorghum is a name given to varieties of a species of sorghum: *Sorghum bicolor*. This crop has been cultivated on a small scale in the past for production of table syrup, but other varieties can be grown for production of sugar. The most common types of sorghum species are those used for production of grain.

There are two advantages of sweet sorghum over sugarcane: its great tolerance to a wide range of climatic and soil conditions, and its relatively high yield of ethanol per acre. In addition, the plant can be harvested in three ways: (1) the whole plant can be harvested and stored in its entirety; (2) it can be cut into short lengths (about 4 inches long) when juice extraction is carried out immediately; and (3) it can be harvested and chopped for ensilage. Since many varieties of sweet sorghum bear significant quantities of grain (milo), the harvesting procedure will have to take this fact into account.

The leaves and fibrous residue of sweet sorghum contain large quantities of protein, making the residue from the extraction of juice or from fermentation a valuable livestock feed. The fibrous residue can also be used as boiler feed.

Jerusalem Artichokes. The Jerusalem artichoke has shown excellent potential as an alternative sugar crop. A member of the sunflower family, this crop is native to North America and well-adapted to northern climates [2]. Like the sugar beet, the Jerusalem artichoke produces sugar in the top growth and stores it in the roots and tuber. It can grow in a variety of soils, and it is not demanding of soil fertility. The Jerusalem artichoke is a perennial; small tubers left in the field will produce the next season's crop, so no plowing or seeding is necessary.

Although the Jerusalem artichoke traditionally has been grown for the tuber, an alternative to harvesting the tuber does exist. It has been noted that the majority of the sugar produced in the leaves does not enter the tuber until the plant has nearly reached the end of its productive life [3]. Thus, it may be possible to harvest the Jerusalem artichoke when the sugar content in the stalk reaches a maximum, thereby avoiding harvesting the tuber. In this case, the harvesting equipment and procedures are essentially the same as for harvesting sweet sorghum or corn for ensilage.

Fodder Beets. Another promising sugar crop which is presently being developed in New Zealand is the fodder beet. The fodder beet is a high yielding forage crop obtained by crossing two other beet species, sugar beets and mangolds. It is similar in most agronomic respects to sugar beets. The attraction of this crop lies in its higher yield of fermentable sugars per acre relative to sugar beets and its comparatively high resistance to loss of fermentable sugars during storage [4]. Culture of fodder beets is also less demanding than sugar beets.

Fruit Crops. Fruit crops (e.g., grapes, apricots, peaches, and pears) are another type of feedstock in the sugar crop category. Typically, fruit crops such as grapes are used as the feedstock in wine production. These crops are not likely to be used as feedstocks for production of fuel-grade ethanol because of their high market value for direct human consumption. However, the coproducts of processing fruit crops are likely to be used as feedstocks because fermentation is an economical method for reducing the potential environmental impact of untreated wastes containing fermentable sugars.

Starch Crops

In starch crops, most of the six-carbon sugar units are linked together in long, branched chains (called starch). Yeast cannot use these chains to produce ethanol. The starch chains must be broken down into individual six-

Corn is one of the Most Popular Ethanol Feedstocks, in Part Due to its Relatively Low Cost of Production

carbon units or groups of two units. The starch conversion process, described in the previous chapter, is relatively simple because the bonds in the starch chain can be broken in an inexpensive manner by the use of heat and enzymes, or by a mild acid solution.

From the standpoint of ethanol production, the long, branched chain arrangement of six-carbon sugar units in starch crops has advantages and disadvantages. The principal disadvantage is the additional equipment, labor, and energy costs associated with breaking down the chain so that the individual sugar units can be used by the yeast. However, this cost is not very large in relation to all of the other costs involved in ethanol production. The principal advantage in starch crops is the relative ease with which these crops can be stored, with minimal loss of the fermentable portion. Ease of storage is related to the fact that a conversion step is needed prior to fermentation: many microorganisms, including yeast, can utilize individual or small groups of sugar units, but not long chains. Some microorganisms present in the environment produce the enzymes needed to break up the chains, but unless certain conditions (such as moisture, temperature, and pH) are just right, the rate of conversion is very slow. When crops and other feeds are dried to about 12% moisture—the percentage at which most microorganisms cannot survive—the deterioration of starch and other valuable components (for example, protein and fats) is minimal. There are basically two subcategories of starch crops: grains (e.g., corn, sorghum, wheat, and barley) and tubers (e.g., potatoes and sweet potatoes). The production of beverage-grade ethanol from both types of starch crops is a well established practice.

Much of the current agronomic research on optimizing the production of ethanol and livestock feed from agricultural crops is focused on unconventional sugar crops such as sweet sorghum. However, opportunities also exist for selecting new varieties of grains and tubers that produce more ethanol per acre. For example, when selecting a wheat variety, protein content is usually emphasized. However, for ethanol production, high starch content is desired. It is well known that wheat varieties with lower protein content and higher starch content usually produce more grain per acre and, consequently, produce more ethanol per acre.

Crop Residue

The "backbone" of sugar and starch crops—the stalks and leaves—is composed mainly of cellulose. The individual six-carbon sugar units in cellulose are linked together in extremely long chains by a stronger chemical bond than exists in starch. As with starch, cellulose must be broken down into sugar units before it can be used by yeast to make ethanol. However, the breaking of the cellulose bonds is much more complex and costly than the breaking of the starch bonds. Breaking the cellulose into individual sugar units is complicated by the presence of lignin, a complex compound surrounding cellulose, which is even more resistant than cellulose to enzymatic or acidic pretreatment. Because of the high cost of converting liquefied cellulose into fermentable sugars, agricultural residues (as well as other crops having a high percentage of cellulose) are not yet a practical feedstock source for small ethanol plants. Current research may result in feasible cellulosic conversion processes in the future.

Forage Crops

Forage crops (e.g., forage sorghum, Sudan grass) hold promise for ethanol production because, in their early stage of growth, there is very little lignin and the conversion of the cellulose to sugars is more efficient. In addition, the proportion of carbohydrates in the form of cellulose is less than in the mature plant. Since forage crops achieve maximum growth in a relatively short period, they can be harvested as many as four times in one growing season [5]. For this reason, forage crops cut as green chop may have the highest yield of dry material of any storage crop. In addition to cellulose, forage crops contain significant quantities of starch and fermentable sugars which can also be converted to ethanol. The residues from fermentation containing nonfermentable sugars, protein, and other components may be used for livestock feed.

The principal characteristics of the feedstock types considered in this section are summarized in Table IV-1.

COPRODUCT YIELDS

Ethanol

The yield of ethanol from agricultural crops can be

TABLE IV-1. SUMMARY OF FEEDSTOCK CHARACTERISTICS

Type of Feedstock	Processing Needed Prior to Fermentation	Principal Advantage(s)	Principal Disadvantage(s)
Sugar Crops (e.g., sugar beets, sweet sorghum, sugarcane, fodder beet, Jerusalem artichoke)	Milling to extract sugar.	1. Preparation is minimal. 2. High yields of ethanol per acre. 3. Crop coproducts have value as fuel, livestock feed, or soil amendment.	1. Storage may result in loss of sugar. 2. Cultivation practices are not wide-spread, especially with "nonconventional" crops.
Starch Crops: Grains (e.g., corn, wheat, sorghum, barley) Tubers (e.g., potatoes, sweet potatoes)	Milling, liquefaction, and saccharification.	1. Storage techniques are well developed. 2. Cultivation practices are widespread with grains. 3. Livestock coproduct is relatively high in protein.	1. Preparation involves additional equipment, labor, and energy costs. 2. DDG from aflatoxin-contaminated grain is not suitable as animal feed.
Cellulosic: Crop Residues (e.g., corn stover, wheat straw) Forages (e.g., alfalfa, Sudan grass, forage sorghum	Milling and hydrolysis of the linkages.	1. Use involves no integration with the livestock feed market. 2. Availability is widespread.	1. No commercially cost-effective process exists for hydrolysis of the linkages.

estimated if the amount of fermentable components—sugar, starch, and cellulose—is known prior to fermentation. If the yield is predicted based on percentages at the time of harvest, then the loss of fermentable solids during storage must be taken into account. This factor can be significant in the case of sugar crops, as discussed earlier.

The potential yield of ethanol is roughly one-half pound of ethanol for each pound of sugar. However, not all of the carbohydrate is made available to the yeasts as fermentable sugars, nor do the yeasts convert all of the fermentable sugars to ethanol. Thus, for estimating purposes, the yield of ethanol is roughly one gallon for each 15 pounds of sugar or starch in the crop at the time the material is actually fermented. Because of the many variables in the conversion of liquefied cellulose to fermentable sugar, it is difficult to estimate active ethanol yields from cellulose.

Carbon Dioxide

The fermentation of six-carbon sugars by yeast results in the formation of carbon dioxide as well as ethanol. For every pound of ethanol produced, 0.957 pound of carbon dioxide is formed; stated another way, for every 1 gallon of ethanol produced, 6.33 pounds of carbon dioxide are formed. This ratio is fixed; it is derived from the chemical equation:

$$C_6H_{12}O_6 \longrightarrow 2C_2H_5OH + 2CO_2 + heat$$

(glucose)   (ethanol)   (carbon dioxide)

Other Coproducts

The conversion and fermentation of agricultural crops yield products in addition to ethanol and carbon dioxide. For example, even if pure glucose is fermented, some yeast will be grown, and they would represent a coproduct. These coproducts have considerable eco-

Wheat, Like the Other Cereal Grains, Produces High Ethanol Yields and the Chaff can be Burned for Process Heat

Chopped Forage Crops May Represent Significant Ethanol Production Potential as Technology for Their Use Improves

nomic value, but, since they are excellent cultures for microbial contaminants, they may represent a pollutant if dumped onto the land. Therefore, it becomes doubly important that these coproducts be put to good use.

Sugar crops, after the sugar has been extracted, yield plant residues which consist mostly of cellulose, unextracted sugar, and protein. Some of this material can be used as livestock feed, although the quantity and quality will vary widely with the particular crop. If the crop is of low feeding value, it may be used as fuel for the ethanol plant. This is commonplace when sugarcane is the feedstock.

Sweet sorghum may yield significant quantities of grain (milo), and the plant residue is suitable for silage, which is comparable to corn or sorghum silage except that it has a lower energy value for feeding. Sugar beet pulp from the production of sugar has always been used for livestock feed, as have the tops. Jerusalem artichokes, grown in the Soviet Union on a very large scale, are ensiled and fed to cattle, so the plant residue in this case would be suitable for silage. All of these residues can supply significant amounts of protein and roughage to ruminants.

It is evident that all silage production has the potential for the production of significant quantities of ethanol without affecting the present uses or agricultural markets. By planting silage crops of high sugar content and extracting a part of the sugar for the production of ethanol, the ensiled residue satisfies the existing demand for silage.

Starch feedstock consists mostly of grains and, to a smaller extent, root crops such as potatoes (white or sweet). The production of nonfermentable material in these root crops is much less than in grains, and the use of the residue is similar.

In the case of grains, it is commonplace to cook, ferment, and distill a mash containing the whole grain. The nonfermentable portion then appears in the stillage (the liquid drawn off the bottom of the beer column after stripping off the ethanol). About three-quarters of the nonfermentable material is in suspension in the form of solids ranging from very coarse to very fine texture, and the remainder is in solution in the water. The suspended material may be separated from the liquid and dried. The coarser solids, in this case, are distillers' light grains. The soluble portion may be concentrated to a syrup with from 25% to 45% solids, called distillers' solubles. When dried together with the coarser material, the product is called distillers' dark grains. These nonfermentable solids derived from grain are valuable as high-protein supplements for ruminants in particular. However, if very large quantities of grain are fermented, the great quantity supplied may exceed the demand and lower the prices. Fortunately, the potential demand exceeds the present usage as a protein supplement, since feeding experience has shown that these coproducts can substitute for a significant part of the grain. When the liquid stillage is fed either as it comes from the still or somewhat concentrated, it is especially valuable, since it permits the substitution of straw for a significant proportion of the hay (e.g., alfalfa) normally fed to ruminants.

The nonfermentable portion of the grain can also be used as human food. In the wet milling industry, the grain components are normally separated and the oil is extracted. The starch may be processed for a number of uses, or it may be used as feedstock for ethanol production. The gluten (the principal portion of the protein in the grain) may be separated and processed for sale as, for example, vital gluten (from wheat) or corn gluten. As another option, the solids may be sent through the fermenters and the beer still to appear as distillers' grains.

Grain processing as practiced in large plants is not feasible for small plants. However, a simple form of processing to produce human food may be feasible. Wheat can be simply processed to separate the starch from the combined germ, gluten, and fiber. They form a cohesive, doughy mass which has long been used as a base for meat-analogs. This material can also be incorporated into bread dough to enhance its nutritional value by increasing the protein, fiber, and vitamin (germ) content.

Work at the University of Wisconsin has resulted in the development of a simple, practical processing machine that extracts about 60% of the protein from forage crops in the form of a leaf juice [6]. The protein in the juice can be separated in a dry form to be used as a very high quality human food. The fibrous residue is then in good condition to be hydrolyzed to fermentable sugars. Most of the plant sugars are in the leaf juice and, after separation of the protein, are ready for fermentation. Forage crops have the potential for producing large amounts of ethanol per acre together with large amounts of human-food-grade protein. The protein production potential is conservatively 1,000 pounds per acre, equivalent to 140 bushels per acre of 12%-protein wheat [7].

Representative feedstock composition and coproduct yields are given in Table IV-2. Appendix D provides additional information in the table comparing raw materials for ethanol production. As discussed earlier, these data cannot be applied to specific analyses without giving consideration to the variable nature of the composition of the feedstock and the yield per acre of the crop.

TABLE IV-2. REPRESENTATIVE YIELDS OF SOME MAJOR DOMESTIC FEEDSTOCKS

Crop	Ethanol Yield
Cereal grains	2.5 gal/bu
Potatoes	1.4 gal/cwt
Sugar beets	20 gal/ton

AGRONOMIC CONSIDERATIONS

A simple comparison of potential ethanol yield per acre of various crops will not rank the crops in terms of economic value for production of ethanol. The crops vary considerably in their demands on the soil, demands for water, need for fertilization, susceptibility to disease or insect damage, etc. These factors critically influence the economics of producing a crop. Fortunately, forage crops which have the potential for producing large amounts of ethanol per acre have specific agronomic advantages relative to some of the principal grain crops (e.g., corn).

The nonfruiting crops, including forage crops, some varieties of high-sugar sorghum, and Jerusalem artichokes, are less susceptible to catastrophic loss (e.g., due to hail, frost, insects, disease, etc.), and, in fact, are less likely to suffer significant loss of production due to adverse circumstances of any sort than are fruiting crops such as grains. Furthermore, forage crops and Jerusalem artichokes are less demanding in their culture than almost any grain. Their cost of culture is usually lower than for grains on the same farm, and they have great potential for planting on marginal land.

FEEDSTOCK CONSIDERATIONS

It is apparent from the foregoing discussion that the selection of feedstocks for ethanol production will vary from region to region, and even from farm to farm. The results of development work now being carried out will influence choices but, most significantly, the additional choices open to farmers resulting from the opportunity to produce feedstocks for ethanol production from a large variety of crops will alter the patterns of farming. It is not possible to predict what new patterns will evolve. However, it is clear that there will be benefits from the creation of choices in the form of new markets for existing crops and alternative crops for existing markets.

In the near future, ethanol is likely to be produced primarily from grain. However, the development of processes for the effective use of other crops should yield results in the near term which could bring about a rapid increase in the use of nongrain feedstocks.

REFERENCES

1. Nathan, R. A. *Fuels from Sugar Crops.* DOE Critical Review Series. 1978. Available from NTIS, #TID-22781.

2. Stauffer, M. D.; Chubey, B. B.; Dorrell, D. G. *Jerusalem Artichoke.* A publication of Agriculture Canada, Research Station, P. O. Box 3001, Morden, Manitoba, ROG 1JO, Canada. 1975.

3. Incoll, L. D.; Neales, T. F. "The Stem as a Temporary Sink before Tuberization." *Helianthus Tuberosus L. Journal of Experimental Botany 21.* (67); 1970; pp. 469-476.

4. Earl, W. B.; Brown, W. A. "Alcohol Fuels from Biomass in New Zealand: The Energetics and Economics of Production and Processing." *Third International Symposium on Alcohol Fuels Technology.* Vol. I, pp. 1-11. Asilomar, CA; May 28-31, 1979.

5. Linden, J. D.; Hedrick, W. C.; Moreira, A. R.; Smith, D. H.; Villet, R. H. *Enzymatic Hydrolysis of the Lignocellulosic Component from Vegetative Forage Crops*. Paper presented before the Second Symposium on Biotechnology in Energy Production and Conversion; October 3-5, 1979. Available from James C. Linden, Department of Agricultural and Chemical Engineering, Colorado State University, Fort Collins, CO 80523.

6. Besken, K. E.; et al. "Reducing the Energy Requirements of Plant Juice Protein Production."Paper presented at the 1975 Annual Meeting of the American Society of Agricultural Engineers; paper no. 75-1056, 1975.

7. Mann, H. O.; et al. "Yield and Quality—Sudan, Sorghum-Sudan, and Pearl Millet Hybrids." Progress Report, Colorado State University, Fort Collins, CO; 1975.

CHAPTER V
Plant Design

CHAPTER V
PLANT DESIGN

The criteria affecting the decision to produce ethanol and establishing a production facility can be categorized into two groups: fixed and variable. The fixed criteria are basically how much ethanol and coproducts can be produced and sold. These issues were discussed in Chapter II. This chapter is concerned with the second set of criteria and their effect on plant design.

Plant design is delineated through established procedures which are complex and interrelated. The essential elements, however, are described here.

The first step is to define a set of criteria which affect plant design. These criteria (not necessarily in order of importance) are:

- amount of labor that can be dedicated to operating a plant;

- size of initial investment and operating cost that can be managed in relation to the specific financial situation and/or business organization;

- ability to maintain equipment both in terms of time to do it and anticipated expense;

- federal, state, and local regulations on environmental discharges, transportation of product, licensing, etc;

- intended use (on-farm use and/or sales) of chemicals;

- desired form of coproducts;

- safety factors;

- availability and expense of heat source; and

- desired flexibility in operation and feedstocks.

The second step is to relate these criteria to the plant as a whole in order to set up a framework or context for plant operations. The third step is complex and involves relating the individual systems or components of production to this framework and to other connected systems within the plant. Finally, once the major systems have been defined, process control systems can be integrated where necessary. This design process leads to specifying equipment for the individual systems and process control.

After the process is discussed from overall plant considerations through individual system considerations to process control, a representative ethanol plant is described. It is an example to illustrate ethanol production technology and not a state-of-the-art or recommended design.

OVERALL PLANT CONSIDERATIONS

Before individual systems and their resulting equipment specifications are examined, the criteria listed above are examined in relation to the overall plant. This establishes a set of constraints against which individual systems can be correlated.

Required Labor

The expense the operation can bear for labor must be considered. To some extent the latter concern is modified by the size of plant selected (the expense for labor is less per gallon the more gallons produced). If it is possible to accomplish the required tasks within the context of daily farming activities, additional outside labor will not be required. A plant operated primarily by one person should, in general, require attention only twice—or at most three times—a day. If possible, the time required at each visit should not exceed 2 hours. The labor availability directly affects the amount and type of control and instrumentation that the plant requires, but it is not the sole defining criteria for plant specification.

Maintenance

The plant should be relatively easy to maintain and not require extensive expertise or expensive equipment.

Feedstocks

The process should use crop material in the form in which it is usually or most economically stored (e.g., forage crops should be stored as ensilage).

Use

The choice of whether to produce anhydrous or lower-proof ethanol depends upon the intended use or market and may also have seasonal dependencies. Use of lower-proof ethanols in spark-ignition tractors and trucks poses no major problems during summertime (or other

Labor Requirements for Ethanol Production can be a Part of the Normal Farm Work Routine

periods of moderate ambient temperature). Any engine equipped for dual injection does not require anhydrous ethanol during moderate seasons (or in moderate climates). If the ethanol is to be sold to blenders for use as gasohol, the capability to produce anhydrous ethanol may be mandatory.

Heat Source

Agricultural residues, coal, waste wood, municipal waste, producer gas, geothermal water, solar, and wind are the preferred possibilities for heat sources. Examples of these considerations are shown in Table V-1. Each poses separate requirements on the boiler selected, the type and amount of instrumentation necessary to fulfill tending (labor) criteria, and the cash flow necessary to purchase the necessary quantity (if not produced on-farm). This last consideration is modified by approaches that minimize the total plant energy demand.

Safety

An ethanol plant poses several specific hazards. Some of these are enumerated in Table V-2 along with options for properly addressing them.

Coproduct Form and Generation

Sale or use of the coproducts of ethanol production is an important factor in overall profitability. Markets must be carefully weighed to assure that competitive influences do not diminish the value of the coproduct that results from the selected system. In some areas, it is conceivable that the local demand can consume the coproduct produced by many closely located small plants; in other areas, the local market may only be able to absorb the coproducts from one plant. If the latter situation occurs, this either depresses the local coproduct market value or encourages the purchase of equipment to modify coproduct form or type so that it can be transported to different markets.

Flexibility in Operation and Feedstocks

Plant profitability should not hinge on the basis of theoretical maximum capacity. Over a period of time, any of a myriad of unforeseen possibilities can interrupt operations and depress yields. Market (or redundant commodity) variables or farm operation considerations may indicate a need to switch feedstocks. Therefore, the equipment for preparation and conversion should be capable of handling cereal grain and at least one of the following:

- ensiled forage material;

- starchy roots and tubers; or

- sugar beets, or other storable, high-sugar-content plant parts.

Compliance with Environmental Regulations and Guidelines

Liquid and gaseous effluents should be handled in compliance with appropriate regulations and standards.

Initial Investment and Operating Costs

All of the preceding criteria impact capital or operating costs. Each criterion can influence production rates

TABLE V-1. HEAT SOURCE SELECTION CONSIDERATIONS

Heat Source	Heating Value (dry basis)	Form	Special Equipment Req'd	Boiler Types	Source	Particular Advantages	Particular Disadvantages
Agriculture Residuals	3,000–8,000 Btu/lb	Solid	Handling and feeding eqpmt.; collection eqpmt.	Batch burner– fire tube; fluidized bed	Farm	Inexpensive; produced on-farm	Low bulk density; requires very large storage area
Coal	9,000–12,000 Btu/lb	Solid	High sulfur coal requires stack scrubber	Conventional grate– fire tube; fluidized bed	Mines	Widely available demonstrated technology for combustion	Potentially expensive; no assured availability; pollution problems
Waste Wood	5,000–12,000 Btu/lb	Solid	Chipper or log feeder	Conventional fluidized bed	Forests	Clean burning; inexpensive where available	Not uniformly available
Municipal Solid Waste	8,000 Btu/lb	Solid	Sorting eqpmt.	Fluidized bed or conventional fire tube	Cities	Inexpensive	Not widely available in rural areas
Pyrolysis Gas		Gas	Pyrolyzer– fluidized bed	Conventional gas-boiler	Carbonaceous materials	Can use conventional gas-fired boilers	Requires additional piece of equipment
Geothermal	N.A.	Steam/ hot water	Heat exchanger	Heat exchanger water tube	Geothermal source	Fuel cost is zero	Capital costs for well and heat exchanger can be extremely high
Solar	N.A.	Radiation	Collectors, concentrators, storage batteries, or systems	Water tube	Sun	Fuel cost is zero	Capital costs can be high for required equipment
Wind	N.A.	Kinetic energy	Turbines, storage batteries, or systems	Electric	Indirect solar	Fuel cost is zero	Capital costs can be high for required equipment

which, in turn, change the income potential of the plant. An optimum investment situation is reached only through repeated iterations to balance equipment requirements against cost in order to achieve favorable earnings.

INDIVIDUAL SYSTEM CONSIDERATIONS

Design considerations define separate specific jobs which require different tools or equipment. Each step depends upon the criteria involved and influences related steps. Each of the components and systems of the plant must be examined with respect to these criteria Figure V-1 diagrams anhydrous ethanol production. The typical plant that produces anhydrous ethanol contains the following systems and/or components: feedstock handling and storage, conversion of carbohydrates to simple sugars, fermentation, distillation, drying ethanol, and stillage processing.

Feedstock Handling and Storage

Grain. A small plant should be able to use cereal grains. Since grains are commonly stored on farms in large quantity, and since grain-growing farms have the basic equipment for moving the grain out of storage, handling should not be excessively time-consuming. The increasing popularity of storing grain at high moisture content provides advantages since harvesting can be done earlier and grain drying can be avoided. When stored as whole grain, the handling requirements are identical to those of dry grain. If the grain is ground and stored in a bunker, the handling involves additional labor since it must be removed from the bunker and loaded into a grainery from which it can be fed by an auger into the cooker. This operation probably could be performed once each week, so the grains need not be ground daily as with whole grain.

Roots and tubers. Potatoes, sugar beets, fodder beets, and Jerusalem artichokes are generally stored whole in cool, dry locations to inhibit spontaneous fermentation by the bacteria present. The juice from the last three can be extracted but it can only be stored for long periods of time at very high sugar concentrations. This requires expensive evaporation equipment and large storage tanks.

Equipment for Handling and Storage of Crop Residues is Currently Available from Farm Equipment Manufacturers

TABLE V-2. ETHANOL PLANT HAZARDS

Hazards	Precautions
1. Overpressurization; explosion of boiler	• Regularly maintained/checked safety boiler "pop" valves set to relieve when pressure exceeds the maximum safe pressure of the boiler or delivery lines. • Strict adherence to boiler manufacturer's operating procedure. • If boiler pressure exceeds 20 psi, acquire ASME boiler operator certification. Continuous operator attendance required during boiler operation.
2. Scalding from steam gasket leaks	• Place baffles around flanges to direct steam jets away from operating areas. • (Option) Use welded joints in all steam delivery lines.
3. Contact burns from steam lines	• Insulate all steam delivery lines.
4. Ignition of ethanol leaks/fumes or grain dust	• If electric pump motors are used, use fully enclosed explosion-proof motors. • (Option) Use hydraulic pump drives; main hydraulic pump and reservoir should be physically isolated from ethanol tanks, dehydration section, distillation columns, condenser. • Fully ground all equipment to prevent static electricity build-up. • Never smoke or strike matches around ethanol tanks, dehydration section, distillation columns, condenser. • Never use metal grinders, cutting torches, welders, etc. around systems or equipment containing ethanol. Flush and vent all vessels prior to performing any of these operations.
5. Handling acids/bases	• Never breathe the fumes of concentrated acids or bases. • Never store concentrated acids in carbon steel containers. • Mix or dilute acids and bases slowly—allow heat of mixing to dissipate. • Immediately flush skin exposed to acid or base with copious quantities of water.

TABLE V-2. ETHANOL PLANT HAZARDS—*Continued*

Hazards	Precautions
	• Wear goggles whenever handling concentrated acids or bases; flush eyes with water and immediately call physician if any gets in eyes.
	• Do not store acids or bases overhead work areas or equipment.
	• Do not carry acids or bases in open buckets.
	• Select proper materials of construction for all acid or base storage containers, delivery aides, valves, etc.
6. Suffocation	• Never enter the fermenters, beer well, or stillage tank unless they are properly vented.

Belt conveyers will suffice for handling these root crops and tubers. Cleaning equipment should be provided to prevent dirt and rocks from building up in the fermentation plant.

Sugar Crops. Stalks from sugarcane, sweet sorghum, and Jerusalem artichokes cannot be stored for long periods of time at high moisture content. Drying generally causes some loss of sugar. Field drying has not been successful in warm climates for sugarcane and sweet sorghum. Work is being conducted in field drying for sweet sorghum in cooler climates; results are encouraging though no conclusions can be drawn yet.

Canes or stalks are generally baled and the cut ends and cuts from leaf stripping are seared to prevent loss of juice.

A large volume of material is required to produce a relatively small amount of sugar, thus a large amount of storage space is necessary. Handling is accomplished with loaders or bale movers.

Conversion of Carbohydrates to Simple Sugars

Processing options available for converting carbohydrates to simple sugars are:

• enzymatic versus acid hydrolysis;

• high-temperature versus low-temperature cooking;

• continuous versus batch processing; and

• separation versus nonseparation of fermentable nonsolids.

Enzymatic versus acid hydrolysis. Enzymatic hydrolysis of the starch to sugar is carried out while cooling the cooked meal to fermentation temperature. The saccharifying enzyme is added at about 130° F, and this temperature is maintained for about 30 minutes to allow nearly complete hydrolysis following which the mash is cooled to fermentation temperature. A high-activity enzyme is added prior to cooking so that the starch is quickly converted to soluble polymeric sugars. The saccharifying enzyme reduces these sugars to monomeric sugars. Temperature and pH must be controlled within specific limits or enzyme activity decreases and cooking time is lengthened. Thus the

Crops for Ethanol Production fit Well into Normal Rotation Practices

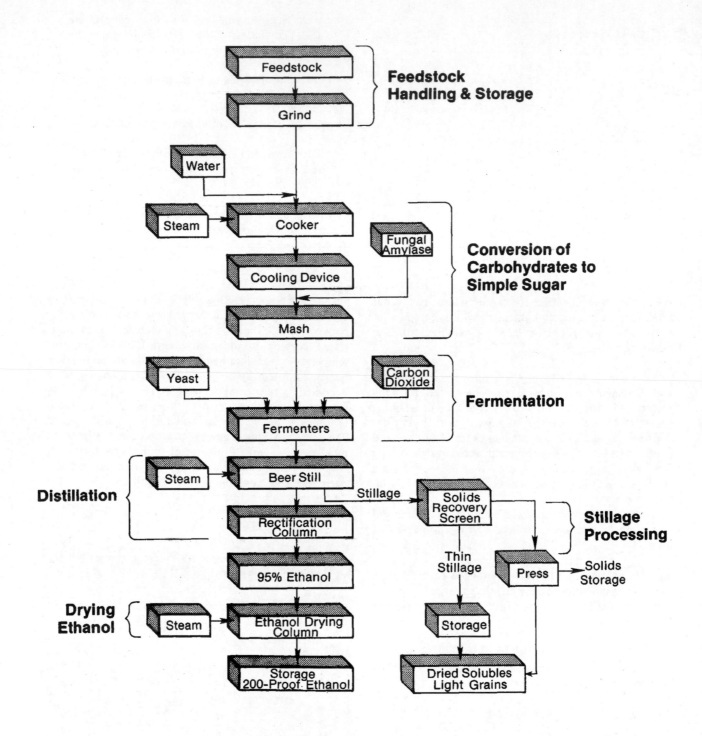

Figure V-1. Anhydrous Ethanol Production Flow Chart

equipment for heating and cooling and the addition of acid or base are necessary.

Acid hydrolysis of starch is accomplished by directly contacting starch with dilute acid to break the polymer bonds. This process hydrolyzes the starch very rapidly at cooking temperatures and reduces the time needed for cooking. Since the resulting pH is lower than desired for fermentation, it may be increased after fermentation is complete by neutralizing some of the acid with either powdered limestone or ammonium hydroxide. It also may be desirable to add a small amount of gluco-amylase enzyme after pH correction in order to convert the remaining dextrins.

High-temperature versus low-temperature cooking. Grain must be cooked to rupture the starch granules and to make the starch accessible to the hydrolysis agent. Cooking time and temperature are related in an inverse ratio; high temperatures shorten cooking time. Industry practice is to heat the meal-water mixture by injecting steam directly rather than by heat transfer through the wall of the vessel. The latter procedure runs the risk of causing the meal to stick to the wall; the subsequent scorching or burning would necessitate a shutdown to clean the surface.

High-temperature cooking implies a high-pressure boiler. Because regulations may require an operator in constant attendance for a high-pressure boiler operation, the actual production gain attributable to the high temperature must be weighed against the cost of the operator. If there are other supporting rationale for having the operator, the entire cost does not have to be offset by the production gain.

Continuous versus batch processes. Cooking can be accomplished with continuous or batch processes. Batch cooking can be done in the fermenter itself or in a separate vessel. When cooking is done in the fermenter, less pumping is needed and the fermenter is automatically sterilized before fermenting each batch. There is one less vessel, but the fermenters are slightly larger than those used when cooking is done in a separate vessel. It is necessary to have cooling coils and an agitator in each fermenter. If cooking is done in a separate vessel, there are advantages to selecting a continuous cooker. The continuous cooker is smaller than the fermenter, and continuous cooking and hydrolysis lend themselves very well to automatic, unattended operation. Energy consumption is less because it is easier to use counterflow heat exchangers to heat the water for mixing the meal while cooling the cooked meal. The load on the boiler with a continuous cooker is constant. Constant boiler load can be achieved with a batch cooker by having a separate vessel for preheating the water, but this increases the cost when using enzymes.

Continuous cooking offers a high-speed, high-yield choice that does not require constant attention. Cooking at atmospheric pressure with a temperature a little over 200° F yields a good conversion ratio of starch to sugar, and no high-pressure piping or pumps are required.

Separation versus nonseparation of nonfermentable solids. The hydrolyzed mash contains solids and dissolved proteins as well as sugar. There are some advantages to separating the solids before fermenting the mash, and such a step is necessary for continuous fermentation. Batch fermentation requires separation of the solids if the yeast is to be recycled. If the solids are separated at this point, the beer column will require cleaning much less frequently, thus increasing the feasibility of a packed beer column rather than plates. The sugars that cling to the solids are removed with the solids. If not recovered, the sugar contained on the solids would represent a loss of 20% of the ethanol. Washing the solids with the mash water is a way of recovering most of the sugar.

Fermentation

Continuous fermentation. The advantage of continuous fermentation of clarified beer is the ability to use high concentrations of yeast (this is possible because the yeast does not leave the fermenter). The high concentration of yeast results in rapid fermentation and, correspondingly, a smaller fermenter can be used. However, infection with undesired microorganisms can be troublesome because large volumes of mash can be ruined before the problem becomes apparent.

Batch fermentation. Fermentation time periods similar to those possible with continuous processes can be attained by using high concentrations of yeast in batch fermentation. The high yeast concentrations are economically feasible when the yeast is recycled. Batch fermentations of unclarified mash are routinely accomplished in less than 30 hours. High conversion efficiency is attained as sugar is converted to 10%-alcohol beer without yeast recycle. Further reductions in fermentation require very large quantities of yeast. The increases attained in ethanol production must be weighed against the additional costs of the equipment and time to culture large yeast populations for inoculation.

Specifications of the fermentation tank. The configuration of the fermentation tank has very little influence on system performance. In general, the proportions of the tank should not be extreme. Commonly, tanks are upright cylinders with the height somewhat greater than the diameter. The bottom may be flat (but sloped for drainage) or conical. The construction materials may be carbon steel (commonplace), stainless steel, copper, wood, fiberglass, reinforced plastic, or concrete coated on the inside with sprayed-on vinyl. Usually, the tanks are covered to permit collection of the CO_2 evolved during fermentation so that the ethanol which evaporates with it can be recovered.

Many potential feedstocks are characterized by relatively large amounts of fibrous material. Fermentation of sugar-rich material such as sugar beets, sweet sorghum, Jerusalem artichokes, and sugarcane as chips is not a demonstrated technology and it has many inherent problems. Typically, the weight of the nonfermentable solids is equal or somewhat greater than the weight of fermentable material. This is in contrast to grain mashers which contain roughly twice as much fermentable material as nonfermentable material in the mash. The volume occupied by the nonfermentable solids reduces the effective capacity of the fermenter. This means that larger fermenters must be constructed to equal the production rates from grain fermenters. Furthermore, the high volume of nonfermentable material limits sugar concentrations and, hence, the beer produced is generally lower in concentration (6% versus 10%) than that obtained from grain mashes. This fact increases the energy spent in distillation.

Since the nonfermentable solid chips are of larger size, it is unlikely that the beer containing the solids could be run through the beer column. It may be necessary to separate the solids from the beer after fermentation because of the potential for plugging the still. The separation can be easily accomplished, but a significant proportion of the ethanol (about 20%) would be carried away by the dewatering solids. If recovery is attempted by "washing out," the ethanol will be much more dilute than the beer. Since much less water is added to these feedstocks than to grain (the feedstock contains large amounts of water), only part of the dilute ethanol solution from the washing out can be recycled through the fermenter. The rest would be mixed with the beer, reducing the concentration of ethanol in the beer which, in turn, increases the energy required for distillation. Another approach is to evaporate the ethanol from the residue. By indirectly heating the residue, the resulting ethanol-water vapor mixture can be introduced into the beer column at the appropriate point. This results in a slight increase in energy consumption for distillation.

The fermenter for high-bulk feedstocks differs somewhat from that used for mash. The large volume of insoluble residue increases the demands on the removal pump and pipe plugging is more probable. Agitators must be sized to be self-cleaning and must prevent massive settling. High-speed and high-power agitators must be used to accomplish this.

The equipment for separating the fibrous residue from the beer when fermenting sugar crops could be used also to clarify the grain mash prior to fermentation. This would make possible yeast recyle in batch fermentation of grain.

Temperature control. Since there is some heat generated during fermentation, care must be taken to ensure that the temperature does not rise too high and kill the yeast. In fermenters the size of those for on-farm plants, the heat loss through the metal fermenter walls is sufficient to keep the temperature from rising too high when the outside air is cooler than the fermenter. Active cooling must be provided during the periods when the temperature differential cannot remove the heat that is generated. The maximum heat generation and heat loss must be estimated for the particular fermenter to assure that water cooling provisions are adequate.

Distillation

Preheater. The beer is preheated by the hot stillage from the bottom of the beer column before being introduced into the top of the beer column. This requires a heat exchanger. The stillage is acidic and hot so copper or stainless steel tubing should be used to minimize corrosion to ensure a reasonable life. Because the solids are proteinaceous, the same protein build-up that plugs the beer still over a period of time can be expected on the stillage side of the heat exchanger. This mandates accessibility for cleaning.

Beer column requirements. The beer column must accept a beer with a high solids content if the beer is not clarified. Not only are there solids in suspension, but also some of the protein tends to build up a rather rubbery coating on all internal surfaces. Plate columns offer the advantage of relatively greater cleaning ease when compared to packed columns. Even if the beer is clarified, there will be a gradual build-up of protein on the inner surfaces. This coating must be removed periodically. If the plates can be removed easily, this cleaning may be done outside of the column. Otherwise, a caustic solution run through the column will clean it.

The relatively low pH and high temperature of the beer column will corrode mild steel internals, and the use of stainless steel or copper will greatly prolong the life expectancy of the plates in particular. Nevertheless, many on-farm plants are being constructed with mild steel plates and columns in the interest of low first cost and ease of fabrication with limited shop equipment. Only experience will indicate the life expectancy of mild steel beer columns.

Introducing steam into the bottom of the beer column rather than condensing steam in an indirect heat exchanger in the base of the column is a common practice. The latter procedure is inherently less efficient but does not increase the total volume of water in the stillage as does the former. Indirect heating coils also tend to suffer from scale buildup.

Rectifying column. The rectifying column does not have to handle liquids with high solids content and there is no protein buildup, thus a packed column suffers no inherent disadvantage and enjoys the advantage in operating stability. The packing can be a noncorroding material such as ceramic or glass.

General considerations. Plate spacing in the large columns of commercial distilleries is large enough to permit access to clean the column. The small columns of on-farm plants do not require such large spacing. The shorter columns can be installed in farm buildings of standard eave height and are much easier to work on.

All items of equipment and lines which are at a significantly higher temperature than ambient should be insulated, including the preheated beer line, the columns, the stillage line, etc. Such insulation is more significant for energy conservation in small plants than for large plants.

Drying Ethanol

Addition of a third liquid to the azeotrope. Ethanol can be dehydrated by adding a third liquid such as gasoline to the 190-proof constant boiling azeotrope. This liquid changes the boiling characteristics of the mixture and further separation to anhydrous ethanol can be accomplished in a reflux still. Benzene is used in industry as a third liquid, but it is very hazardous for on-farm use. Gasoline is a suitable alternative liquid and does not pose the same health hazards as benzene, but it fractionates in a distillation column because gasoline is a mixture of many organic substances. This is potentially an expensive way to break the azeotrope unless the internal reflux is very high, thereby minimizing the loss of gasoline from the column. Whatever is choosen for the third liquid, it is basically recirculated continually in the reflux section of the drying column, and thus only very small fractions of makeup are required. The additional expense for equipment and energy must be weighed carefully against alternative drying methods or product value in uses that do not require anhydrous ethanol.

Molecular sieve. The removal of the final 4% to 6% water has also been accomplished on a limited basis using a desiccant (such as synthetic zeolite) commonly known as a molecular sieve. A molecular sieve selectively absorbs water because the pores of the material are smaller than the ethanol molecules but larger than the water molecules. The sieve material is packed into two columns. The ethanol—in either vapor or liquid form—is passed through one column until the material in that column can no longer absorb water. Then the flow is switched to the second column, while hot (450° F) and preferably nonoxidizing gas is passed through the first column to evaporate the water. Carbon dioxide from the fermenters would be suitable for this. Then the flow is automatically switched back to the other column. The total energy requirement for regeneration may be significant (the heat of absorption for some synthetic zeolites is as high as 2,500 Btu/lb). Sieve material is available from the molecular sieve manufacturers listed in Appendix E, but columns of the size required must be fabricated. The molecular sieve material will probably serve for 2,000 cycles or more before significant deterioration occurs.

Selective absorption. Another very promising (though undemonstrated) approach to dehydration of ethanol has been suggested by Ladisch [1]. Various forms of starch (including cracked corn) and cellulose selectively absorb water from ethanol-water vapor. In the case of grains, this opens the possibility that the feedstock could be used to dehydrate the ethanol and, consequently, regeneration would not be required. More investigation and development of this approach is needed.

Stillage Processing

The stillage can be a valuable coproduct of ethanol production. The stillage from cereal grains can be used as a high-protein component in animal feed rations, particularly for ruminants such as steers or dairy cows. Small on-farm plants may be able to directly use the whole stillage as it is produced since the number of cattle needed to consume the stillage is not large (about one head per gallon of ethanol production per day).

Solids separation. The solids can be separated from the water to reduce volume (and hence shipping charges) and to increase storage life. Because the solids contain residual sugars, microbial contaminants rapidly spoil stillage if it is stored wet in warm surroundings. The separation of the solids can be done easily by flowing the stillage over an inclined, curved screen consisting of a number of closely-spaced transverse bars. The solids slide down the surface of the screen, and the liquid flows through the spaces between the bars. The solids come off the screen with about 85% water content, dripping wet. They can drop off the screen into the hopper of a dewatering press which they leave at about 65% water content. Although the solids are still damp, no more water can be easily extracted. The liquid from the screen and dewatering press contains a significant proportion of dissolved proteins and carbohydrates.

Transporting solids. The liquid from the screen and dewatering press still contains a significant proportion of dissolved proteins and carbohydrates. If these damp solids are packed in airtight containers in CO_2 atmosphere, they may be shipped moderate distances and stored for a short time before microbes cause major spoilage. This treatment would enable the solids from most small plants to reach an adequate market. While the solids may easily be separated and dewatered, concentrating the liquid (thin stillage) is not simple. It can be concentrated by evaporation, but the energy consumption is high unless multiple-effect evaporators are used. These evaporators are large and expensive, and may need careful management with such proteinaceous liquids as thin stillage.

Stillage from aflatoxin-contaminated grains or those treated with antibiotics are prohibited from use as animal feed.

Distillers' solubles, which is the low-concentration (3% to 4% solids) solution remaining after the solids are dewatered, must be concentrated to a syrup of about 25% solids before it can be economically shipped moderate distances or stored for short times. In this form it can be sold as a liquid protein to be used in mixed feed or it can be dried along with the damp distillers' grains.

Disposing of thin stillage. If the distance from markets for the ethanol coproduct necessitates separating and dewatering the stillage from an on-farm plant, and if the concentration of the stillage for shipment is not feasible, then the thin stillage must be processed so that it will not be a pollutant when discharged. Thin stillage can be anaerobically fermented to produce methane. Conventional flow-through type digesters are dependent upon so many variables that they cannot be considered commercially feasible for on-farm use. Experimental work with packed-bed digesters is encouraging because of the inherent stability observed [1].

Another way to dispose of the thin stillage is to apply it to the soil with a sprinkler irrigation system. Trials are necessary to evaluate the various processes for handling the thin stillage. Because the stillage is acidic, care must be taken to assure that soil acidity is not adversely affected by this procedure.

PROCESS CONTROL

Smooth, stable, and trouble-free operation of the whole plant is essential to efficient conversion of the crop material. Such operation is, perhaps, more important to the small ethanol plant than to a larger plant, because the latter can achieve efficiency by dependence on powerful control systems and constant attention from skilled operators. Process control begins with equipment characteristics and the integration of equipment. There is an effect on every part of the process if the conditions are changed at any point. A good design will minimize negative effects of such interactions and will prevent any negative disturbance in the system from growing. Noncontinuous processes (e.g., batch fermentation) tend to minimize interactions and to block such disturbances. The basic components requiring process control in a small-scale ethanol plant are cooking and hydrolysis, fermentation, distillation, ethanol drying system pumps and drives, and heat source.

Control of Cooking and Hydrolysis

Input control. All inputs to the process must be controlled closely enough so that the departures from the desired values have inconsequential effects. The batch process has inherently wider tolerances than the continuous process. Tolerances on the grain-water ratio can be fairly loose. A variation of ½% in ethanol content will not seriously disturb the system. This corresponds to about a 3% tolerance on weight or volume measure. Meal measurement should be made by weight, since the weight of meal filling a measured volume will be sensitive to many things, such as grain moisture content, atmospheric humidity, etc. Volume measurement of water is quite accurate and easier than weighing. Similarly, volume measurement of enzymes in liquid form is within system tolerances. Powdered enzymes ideally should be measured by weight but, in fact, the tolerance on the proportion of the enzymes is broad enough so that volume measure also is adequate.

Temperature, pH, and enzyme control. The temperature, pH, and enzyme addition must also be controlled. The allowed variation of several degrees means that measurement of temperature to a more than adequate precision can be easily accomplished with calibrated, fast-response indicators and read-outs. The time dependence brings in other factors for volume and mass. A temperature measurement should be representative of the whole volume of the cooker; however, this may not be possible because, as the whole mass is heating, not all parts are receiving the same heat input at a given moment since some parts are physically far removed from the heat source. This affects not only the accuracy of the temperature reading but also the cooking time and the action of the enzyme. Uniformity of temperature and of enzyme concentration throughout the mass of cooking mash is desired and may be attained by mixing the mass at a high rate. Thus, agitation is needed for the cooker. The temperature during the specific phases of cooking and hydrolysis must be controlled by regulating steam and cooling water flow-rates based on temperature set-points.

Automatic controls. An automatic controller could be used to turn steam and cooling water on and off. The flow of meal, water, enzymes, and yeast could be turned on and off by the same device. Therefore, the loading and preparation of a batch cooker or fermenter could easily be carried out automatically. Safety can be ensured by measuring limiting values of such quantities as temperature, water level, pH, etc., and shutting down the process if these were not satisfied. Any commercial boiler used in a small plant would be equipped with simple, automatic controls including automatic shutdown in case certain conditions are not met. There is a need for an operator to check on the system to assure that nothing goes wrong. For example, the mash can set up during cooking, and it is better to have an operator exercise judgment in this case than to leave it entirely to the controls. Since cooking is the step in which there is the greatest probability of something going wrong, an operator should be present during the early, critical stages of batch cooking. If continuous cooking is used, unattended operation requires that the process be well enough controlled so that there is a small probability of problems arising.

Control of Fermentation

Temperature and pH control. Batch fermentation does not need direct feedback control except to maintain temperature as long as the initial conditions are within acceptable limits. For the small plant, these limits are not very tight. The most significant factors are pH and temperature. Of the two, temperature is most critical. It is very unlikely that the change in pH will be great enough to seriously affect the capacity of the yeast to convert the sugar. Fermentation generates some heat, so the temperature of the fermenter tends to rise. Active cooling must be available to assure that summertime operations are not drastically slowed because of high-temperature yeast retardation.

The temperature of the fermenter can be measured and, if the upper limit is exceeded, cooling can be initiated. It is possible to achieve continuous control of the fermenter temperature through modulation of the cooling rate of the contents. Such a provision may be necessary for very fast fermentation.

Automatic control. Continuous fermentation, like continuous cooking, should have continuous, automatic control if constant attendance by an operator is to be avoided.

The feasibility of continuous, unattended fermentation in on-farm plants has not been demonstrated, although it is a real possibility.

Control with attention at intervals only. The feasibility of batch fermentation with attention at intervals has been established. After initiating the cooking and hydrolysis steps, the operator could evaluate the progress of fermentation at the end of the primary phase and make any adjustments necessary to assure successful completion of the fermentation. This interval between the points requiring operator attention can vary widely, but is usually from 8 to 12 hours. Fermentation can be very fast—as short as 6 hours—but the conditions and procedures for reliably carrying out such fast fermentations have not yet been completely identified and demonstrated. The schedule for attending the plant should allow about 15% additional time over that expected for completion of the fermentation process. This permits the operator to maintain a routine in spite of inevitable variations in fermentation time.

Controls for Distillation

The distillation process lends itself well to unattended operation. Continuous control is not mandatory because the inputs to the columns can easily be established and maintained essentially constant. These inputs include the flow rate of beer, the flow rate of steam, and the reflux flow rate. These are the only independent variables. Many other factors influence column operation, but they are fixed by geometry or are effectively constant. Once the distillation system is stabilized, only changes in ambient temperature might affect the flow balance as long as the beer is of constant ethanol content. Sensitivity to ambient temperature can be minimized by the use of insulation on all elements of the distillation equipment, and by installing the equipment in an insulated building. Occasional operator attention will suffice to correct the inevitable slow drift away from set values. The system also must be adjusted for changes in ethanol content from batch to batch.

Distillation column design can aid in achieving stable operation. Packed columns are somewhat more stable than plate columns, particularly as compared to simple sieve plates.

Starting up the distillation system after shutdown is not difficult and can be accomplished either manually or automatically. An actual sequence of events is portrayed in the representative plant described at the end of this chapter. The process is quite insensitive to the rate of change of inputs, so the demands made on the operator are not great. It is important that the proper sequence be followed and that the operator know what settings are desired for steady-state operation.

Control of Ethanol Drying System

Operation of a molecular sieve is a batch process. As such, it depends on the capacity of the desiccant to ensure completion of drying. No control is necessary except to switch ethanol flow to a regenerated column when the active column becomes water-saturated. Water saturation of the sieve can be detected by a rise in temperature at the discharge of the column. This temperature rise signals the switching of flow to the other column, and regeneration of the inactive column is started immediately. The regeneration gas, probably CO_2 from the fermenter, is heated by flue gas from the boiler. The control consists of initiating flow and setting the temperature. The controller performs two functions: it indicates the flow and sets the temperature of the gas. Two levels of temperature are necessary: the first (about 250° F) is necessary while alcohol clinging to the molecular sieve material is being evaporated; the second (about 450° F) is necessary to evaporate the adsorbed water. Here again, the completion of each phase of the regeneration cycle is signaled by a temperature change at the outlet from the column. Finally, the col-

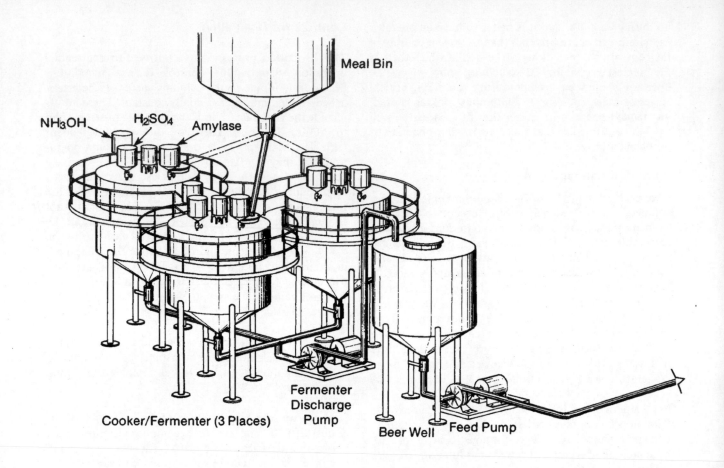

umn is cooled by passing cool CO_2 through it until another outlet temperature change indicates completion of the regeneration cycle. The controls required for a dehydration distillation column are essentially the same as those required for the rectification column.

Controls for Pumps and Drives

The pumps used in this plant can be either centrifugal or any one of a number of forms of positive displacement pumps. The selection of the pump for mash or beer needs to take into account the heavy solids loading (nearly 25% for mash), the low pH (down to 3.5 for the beer), and the mild abrasive action of the mash.

The pumps might be powered by any of a number of different motors. The most probable would be either electric or hydraulic. If electric motors are used, they should be explosion-proof. Constant speed electric motors and pumps are much less expensive than variable-speed motors. Control of the volume flow for the beer pump, the two reflux pumps, and the product pump would involve either throttling with a valve, recirculation of part of the flow through a valve,

or variable-speed pumps. Hydraulic drive permits the installation of the one motor driving the pumps to be located in another part of the building, thereby eliminating a potential ignition source. It also provides inexpensive, reliable, infinitely variable speed control for each motor. Hydraulic drives could also be used for the augers, and the agitators for the cookers and fermenters. Since hydraulics are used universally in farm equipment, their management and maintenance is familar to farmers.

Heat Source Controls

There are basically two processes within the ethanol production system that require heat: the cooking and the distillation steps. Fortunately, this energy can be supplied in low-grade heat (less than 250° F). Potential sources of heat include coal, agriculture residues, solar, wood wastes, municipal wastes, and others. Their physical properties, bulk density, calorific value, moisture content, and chemical constituency vary widely. This, in turn, requires a greater diversity in equipment for handling the fuel and controls for operating the boilers. Agriculture residues vary in bulk

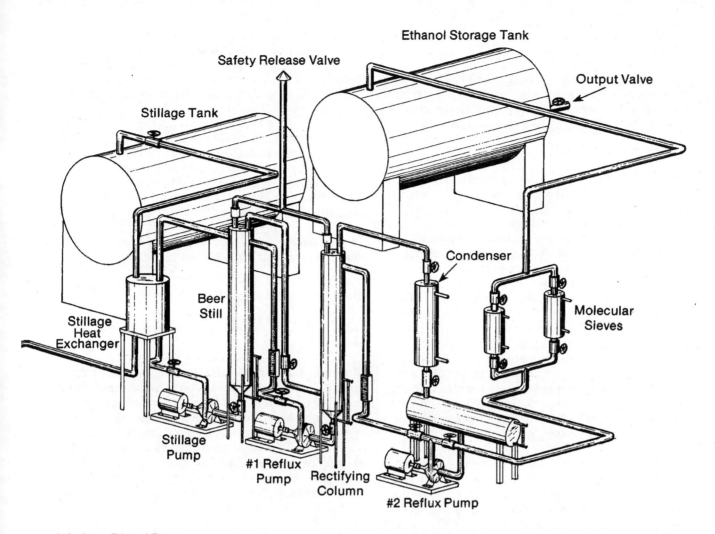

Anhydrous Ethanol Plant

density from 15 to 30 pounds per cubic foot and the calorific value of oven-dry material is generally around 8,000 Btu per pound. This means that a large volume of fuel must be fed to the boiler continuously. For example, a burner has been developed that accepts large, round bales of stover or straw. The boiler feed rate will vary in direct proportion to the demand for steam. This in turn is a function of the distillation rate, the demand for heat for cooking (which varies in relation to the type of cooker and fermenter used—batch or continuous).

Emissions. Emissions controls on the boiler stack are probably minimal, relying instead upon efficient burner operation to minimize particulate emissions. If exhaust gas scrubbers or filters are required equipment, they in turn require feedback control. Filters must be changed on the basis of pressure drop across them which indicates the degree of loading (plugging). Scrubbers require control of liquid flow rate and control of critical chemical parameters.

Boiler safety features. Safety features associated with

the boiler are often connected to the control scheme to protect the boiler from high-pressure rupture and to prevent burnout of the heat expander tubes. Alarm systems can be automated and have devices to alert an operator that attention is needed. For instance, critical control alarms can activate a radio transmitter, or "beeper," that can be worn by the farmer while performing other normal work routine.

REPRESENTATIVE ETHANOL PLANT

General descriptions of major components serve only to define possibilities. In the previous section, considerations for specifying the appropriate equipment to accomplish desired objectives were examined. The following is a description of a specific representative ethanol plant producing ethanol and wet stillage. This representative plant normally produces 25 gallons of anhydrous ethanol per hour. The distillation section can be operated continuously with shutdown as required to remove protein buildup in the beer column. Heat is provided by a boiler that uses agricultural residue as fuel. The plant is designed for maximum flexibility, but its principal feedstocks are cereal grain, with specific emphasis on corn.

This representative plant should not be construed as a best design or the recommended approach. Its primary purpose is to illustrate ethanol production technology.

Overview of the Plant

As shown in Figure V-2, the representative plant has seven main systems: (1) feed preparation and storage, (2) cooker/fermenter, (3) distillation, (4) stillage storage, (5) dehydration, (6) product storage, (7) and boiler. Grain from storage is milled once a week to fill the meal bin. Meal from the bin is mixed to make mash in one of three cooker/fermenters. The three cooker/fermenters operate on a staggered schedule—one starting, one fermenting, and one pumping out—to maintain a full beer well so the distillation section can be run continuously. The beer well provides surge capacity so that the fermenters can be emptied, cleaned, and restarted without having to wait until the still can drain them down. Beer is fed from the beer well to the beer still through a heat exchanger that passes the cool beer counter-current to the hot stillage from the bottom of the beer still. This heats up the beer and recovers some of the heat from the stillage.

The beer still is a sieve-plate column. The feed is introduced at the top of the stripping section. Vapors from the beer column flow into the bottom of the rectifying column where the ethanol fraction is enriched to 95%. The product is condensed and part of it is recycled (refluxed) to the top of the column and if ethanol is being dried at the time, part of it is pumped to the dehydration section. If the ethanol is not being dried, it flows directly to a storage tank for 190-proof ethanol (a separate tank must be used for the anhydrous ethanol).

The stillage that is removed from the bottom of the beer column is pumped through the previously mentioned heat exchanger and is stored in a "whole stillage" tank. This tank provides surge capacity when a truck is unavailable to haul the stillage to the feeder operation.

The distillation columns are designed for inherent stability once flow conditions are established so a minimum of automatic feedback control and instrumentation is required. This not only saves money for this equipment but it reduces instrument and/or controller-related malfunctions. Material flows for cooling and fermentation are initiated manually but proceed automatically. A sequencer microprocessor (a miniature computer) controls temperature and pH in the cooker/fermenters. It also activates addition of enzymes and yeast in the proper amounts at the proper times. At any point, the automatic sequence can be manually overridden.

The period of operation is quite flexible, and allows for interruptions of operation during planting or harvest time. The 25 gallons per hour production is a nominal capacity, not a maximum. All support equipment is similarly sized so that slightly higher production rates can be achieved if desired.

The control and operating logic for the plant is based on minimal requirements for operator attention. Critical activities are performed on a routine periodic basis so that other farming operations can be handled during the bulk of the day. All routines are timed to integrate with normal chore activities without significant disruption.

A complete equipment list is given in Table V-3. The major components are described in Table V-4.

Start-Up and Shutdown

The following is a sequence for starting-up or shutting down the plant.

Preliminaries. For the initial start-up, a yeast culture must be prepared or purchased. The initial yeast culture can use a material such as molasses; later cultures can be grown on recycled stillage. Yeast, molasses, and some water should be added to the yeast culture tank to make the culture. Although yeasts function anaerobically, they propagate aerobically, so some oxygen should be introduced by bubbling a small amount of air through the culture tank. The initial yeast culture will take about 24 hours to mature.

At this time, the boiler can be started. Instructions packaged with the specific boiler will detail necessary steps to bring the unit on-line (essentially the boiler is filled with water and the heat source started). These instructions should be carefully followed, otherwise there is the possibility of explosion.

The next step is the milling of grain for the cooker/fermenter. Enough grain should be milled for two fermentation batches (about 160 bushels).

Prior to loading the fermenter, it should be cleaned well with a strong detergent, rinsed, decontaminated with a strong disinfectant, and then rinsed with cold water to flush out the disinfectant.

Mash Preparation. The amount of meal put in the cooker/fermenter depends upon the size of batch desired. For the first batch it is advisable to be conservative and start small. If the batch is ruined, not as much material is wasted. A 2,000-gallon batch would be a good size for this representative plant. This will require mixing 80 bushels of ground meal with about 500 gallons of water to form a slurry that is about 40% starch.

TABLE V-3. EQUIPMENT FOR REPRESENTATIVE PLANT

Equipment	Description	Equipment	Description
Grain Bin	• ground carbon steel • 360 bu with auger for measuring and loading cooker/fermenter	Heat Exchanger	• 150 ft², tube and shell • copper coil (single tube, 2-in.) diameter) • steel shell
Back-Pressure Regulators	• 0–50 in. of water	Heat Exchanger	• 100 ft² • stack gas • carbon steel
Back-Pressure Regulator	• 100–200 psig	Hydraulic System for Pumps	• with shut-off valves tied to microprocessor monitoring pump pressures and frangible vent temperature
Beer Storage Tank	• 6,000-gal • carbon steel		
Condenser, Distiller	• 225 ft², tube and shell • copper coil (single tube, 1½-ft diameter) • steel shell cooled	Grain Mill	• 300 bu/hr • roller type
Condenser	• 50 ft² • copper coil (single tube) • steel shell	Beer Pump	• positive displacement • hydraulic drive • variable speed • carbon steel, 50 gal/min
Cooker, Fermenter	• 4,500-gal • hydraulic agitator • carbon steel	Yeast Pump	• positive displacement • hydraulic drive • variable speed • carbon steel, 10 gal/min
Microprocessor	• to control heat for cooking, cooling water during fermentation, and addition of enzymes	Feed Pump	• 300 gal/hr • variable speed • positive displacement • hydraulic drive • carbon steel
Beer Still	• 18-ft height • 1-ft diameter • sieve trays • carbon steel	Stillage Pump	• 300 gal/hr • variable speed • positive displacement • hydraulic drive • carbon steel
Alcohol Still	• 24-ft height • 1-ft diameter • sieve trays • carbon steel		
CO_2 Compressor	• 1,500 ft³/hr, 200 psig	Column 2 Bottoms Pump	• 250 gal/hr • open impeller • centrifugal hydraulic drive • carbon steel
Frangibles	• 4–5 psig burst • alarm system • high and low pressure	Column 2 Product and Reflux Pump	• 200 gal/hr

TABLE V-3. EQUIPMENT FOR REPRESENTATIVE PLANT —*Continued*

Equipment	Description	Equipment	Description
	• open impeller • centrifugal hydraulic drive • carbon steel	Pressure Transducers	• 4, 0–100 psig
Ethanol Transfer Pump	• centrifugal • explosion-proof motor • 50 gal/min	Ethanol Drying Columns	• includes molecular sieve packing 3-angstrom synthetic zeolite
Water Pump	• electric • open impeller • centrifugal • 300 gal/min	Condensate Receiver	• 30-gal, horizontal • carbon steel
Rotameter	• water fluid • glass • 25 gal/min	Ethanol Storage Tank	• carbon steel • 9,000-gal
Rotameter	• glass • 0–250 gal/hr	CO_2 Storage	• 100-gal, 200 psig
Rotameter	• glass • 0–150 gal/hr	Stillage Storage Tank	• 4,500-gal • carbon steel
Rotameter	• glass • 0–50 gal/hr	Thermocouples	• type K, stainless sheath
Rotameter–CO_2	• glass • 200 psig • 100 actual ft³/hr	Multichannel Digital Temperature Readout	• 15 channels
Boiler	• 500 hp, with sillage burning system	Ball Valve-65	
		Metering Valve-6	
Stillage Pump	• electric motor • positive displacement • 600 gal/hr	Three-Way Valve-4	
		Snap Valve	
		Water Softener	• 300 gal/hr
Pressure Gauges	• 6, 0–100 psig • 1, 0–200 psig	Yeast Culture Tank	• carbon steel • 200-gal

TABLE V-4. FEATURES OF MAJOR PLANT COMPONENTS

Components	Features	Components	Features
Feedstock Storage and Preparation			
Grain Mill	• roller mill that grinds product to pass a 20-mesh screen	Meal Bin	• corrugated, rolled galvanized steel with 360-bu capacity

TABLE V-4. FEATURES OF MAJOR PLANT COMPONENTS—*Continued*

Components	Features	Components	Features
Auger	• used for feeding meal to cooker/fermenter		• top-mounted feed port
			• hydraulic agitator
Trip buckets	• used to automatically measure meal in proper quantity; as buckets fill, they become unbalanced and tip over into the cooker/fermenter; each time a bucket tips over, it trips a counter; after the desired number of buckets are dumped, the counter automatically shuts off the auger and resets itself to zero		• cooling coils
			• pH meter
			• sodium hydroxide tank
			• dilute sulfuric acid tank
			• temperature-sensing control, preset by sequences
Cooker/Fermenter			• steam injection
3 Cookers	• 4,500-gal right cylinder made of cold-rolled, welded carbon steel	Glucoamylase Enzyme Tanks	• 5-gal capacity
			• fitted with stirrer
			• ball-valve port to cooker/fermenter triggered by sequencer
		Sequencer	• controls cooking fermentation sequences
			• actuates ball-valve to add glucoamylase enzyme after temperature drops from liquefaction step
			• sequences temperature controller for cooker/fermenter
			• sets pH reading for pH controller according to step
			• 6,000-gal capacity
			• cold-rolled, welded carbon steel
			• flat top
			• conical bottom
			• ball-valve port at bottom
			• man-way on top, normally kept closed (used for cleaning access only)

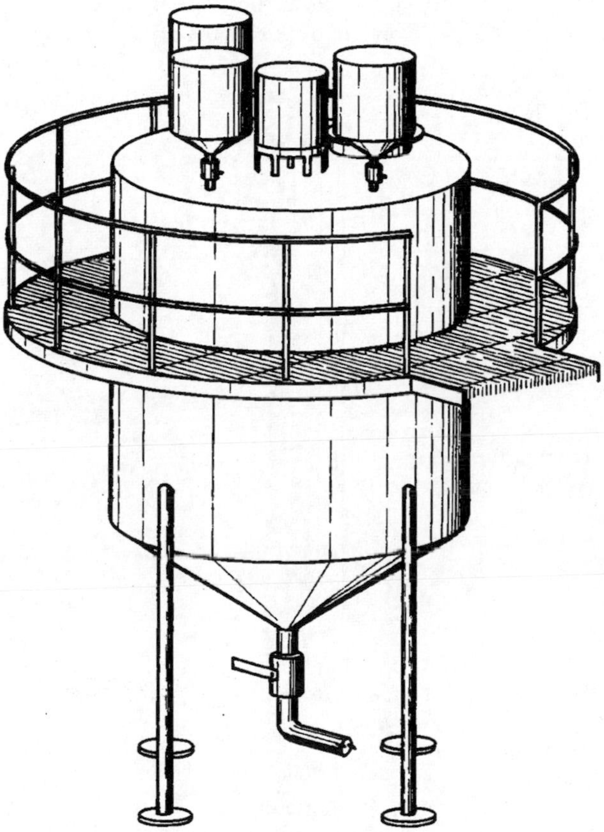

Figure V-3. Cooker/Fermenter

• flat top

• conical bottom

• ball-valve drain port

TABLE V-4. FEATURES OF MAJOR PLANT COMPONENTS—*Continued*

Components	Features	Components	Features
Beer/Stillage Heat Exchanger	• 2-ft diameter, 3-ft tall—beer flows through coil, stillage flows through tank		• steam introduced at bottom through a throttle valve

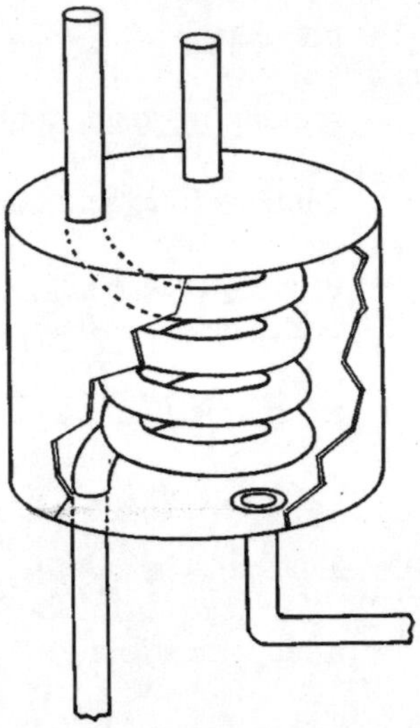

Figure V-4. Beer/Stillage Heat Exchanger

Beer Pump • pump from any of the three cooker/fermenters to beer well, hydraulic motor on pump

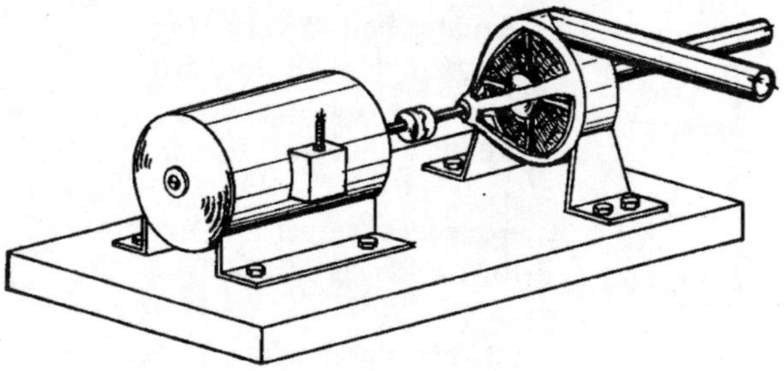

Figure V-5. Beer Pump

Feed Pump • pump beer to distillation system, hydraulic motor on pump

Distillation

Beer Still • 1-ft diameter

• 20-ft tall coated carbon steel pipe with flanged top and bottom

• fitted with a rack of sieve trays that can be removed either through the top or bottom

• steam introduced at bottom through a throttle valve

• pump at the bottom to pump stillage out, hydraulic motor on pump

• input and output flows are controlled through manually adjusted throttle valves

• safety relief valves prevent excess pressure in column

• instrumentation includes temperature indication on feed line and at the bottom of the still, sight-glass on bottom to maintain liquid level, pressure indicators on the outlet of the stillage pump

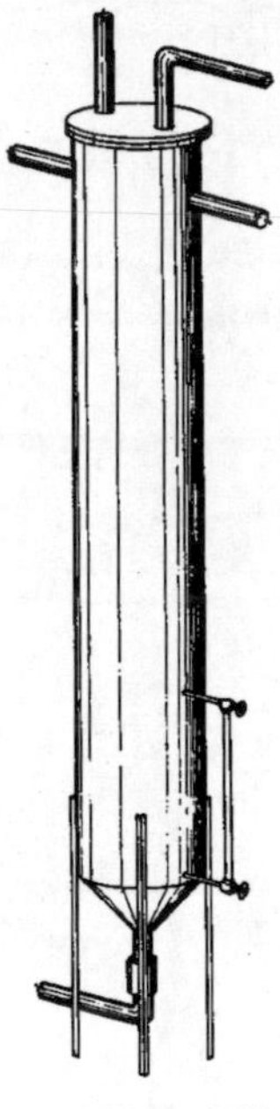

Figure V-6. Beer Still

Rectifying Column

• 20-ft tall
• 1-ft diameter

• coated carbon steel pipe with flanged top, welded bottom to prevent ethanol leaks

• fitted with rack of sieve-plates which can be removed through the top

TABLE V-4. FEATURES OF MAJOR PLANT COMPONENTS—*Continued*

Components	Features	Components	Features

- pump at bottom of column refluxes ethanol at set rate back to beer still, rate is set with throttle valve and rotameter, hydraulic motor on pump

Figure V-7. Rectifying Column Rotameter

- instrumentation consists of temperature indication at top and bottom of column and level indication at bottom by sight-glass, pressure is indicated on the outlet of the recirculation pump

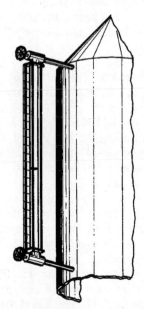

Figure V-8. Rectifying Column Sight-Glass

Condenser
- ethanol condenses (in copper coil), water flows through tank
- cooling water flow-rate is manually adjusted

Dehydration Secton

2 Molecular Sieves
- packed bed
- synthetic zeolite, type 3A-molecular sieve material
- automatic regeneration
- automatic temperature control during regeneration
- throttle flow control to sieves adjusted manually

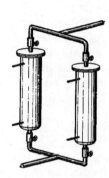

Figure V-9. Molecular Sieves

CO_2 Compressor
- 2-stage air compressor with reservoir (conventional)

Denaturing Tank
- meets Bureau of Alcohol, Tobacco, and Firearms specifications (See Appendix B)

Ethanol Storage

2 Ethanol Storage Tanks
- 3,000-gal capacity each
- same as gasoline storage tanks
- cold-rolled, welded carbon steel

Stillage Storage Tank
- 6,000-gal capacity
- cold-rolled, welded carbon steel

Cooking. The water and meal are blended together as they are added to the cooker/fermenter. It is crucial to use rates that promote mixing and produce no lumps (the agitator should be running). The alpha-amylase enzyme can be blended-in during the mixing (the enzyme must be present and well mixed before the temperature is raised because it is very difficult to disperse the enzyme after gelatinization occurs).

Since cooking in this representative plant is initiated by steam injection during slurry-mixing, the enzyme must be blended in simultaneously. (Dry enzymes should be dispersed in a solution of warm water before mixing is started. This only takes a small amount of water, and the directions come on the package. Liquid enzymes can be added directly.) If the pH is lower than 5.5, it should be adjusted by addition of a calculated amount of sodium hydroxide. If the pH is higher than 7.0, a calculated amount of sulfuric acid should be added. Steam is added at a constant rate to achieve uniform heating. When the temperature reaches 140° F (60° C), the physical characteristics of the mash change noticeably as the slurry of starch becomes a solution of sugar. If there is insufficient enzyme present or if heating is too rapid, a gel will result that is too thick to stir or add additional enzyme to. If a gel does form, more water and enyzme can be added (if there is room in the tank) and the cook can start over.

Once liquefaction occurs, the temperature is uniformly raised to the range for optimum enzyme activity (about 200° F) and held for about half an hour. At the end of this time, a check is made to determine if all of the starch has been converted to sugar. A visual inspection usually is sufficient; incomplete conversion will be indicated by white specks of starch or lumps; a thin, fluid mash indicates good conversion. The mash is held at this temperature until most of the starch is converted to dextrin.

Saccharification. Once the mash is converted to dextrin, the microprocessor is manually started and (1) reduces the temperature of the mash to about 135° F (57° C) by circulating cooling water through the coils; and (2) adds dilute sulfuric acid (H_2SO_4) until the pH drops to between 3.7 and 4.5 (H_2SO_4 addition is controlled by a pH meter and a valve on the H_2SO_4 tank). Once the pH and temperature are within specified ranges, the microprocessor triggers the release of liquid glucoamylase (which must be premixed if dry enzyme is used) from its storage tank. Either sodium hydroxide or sulfuric acid is added automatically as required to maintain proper pH during conversion. The microprocessor also holds the mash at a constant temperature by regulating steam and/or cooling water flow for a preset period of time. The sequencer can be overridden if the conversion is not complete.

Fermentation. After hydrolysis is complete, the sequencer lowers the temperature of the mash to about 85° F by adding the remaining 1,500 gallons of water (and by circulating cooling water thereafter as necessary). The water addition will raise the pH of the solution so the sequencer automatically adjusts the pH to between 4.5 and 5.0. Next, the sequencer adds a premeasured quantity of dispersed distillers' yeast from the yeast tank. (Note that the yeast tank is not on top of the cooker/fermenter as high temperatures during cooking would kill the culture.) Thereafter, the sequencer maintains the temperature between 80° F and 85° F and the pH between 3.0 and 5.0. The agitator speed is reduced from that required during cooking to a rate which prevents solids from settling, but does not disturb the yeast. The batch is then allowed to ferment for 30 to 36 hours.

Pump-Out and Cleanup. After a batch is complete, it is pumped to the beer well and the fermenter is hosed out to remove any remaining solids.

Distillation. Once the beer well is full, the distillation system can be started up. This process involves the following steps.

1. **Turn on the condenser cooling water.**

2. **Purge the still with steam.** This removes oxygen from the system by venting at the top of the second column. When steam is seen coming out of the vent, the steam can be temporarily shut off and the vent closed. Purging the still with steam not only removes oxygen, but also helps to preheat the still.

3. **Pump beer into the still.** The beer is pumped in until it is visible at the top of the sight-glass.

4. **Turn steam on and add beer.** This process of adding beer and watching the liquid level movement to adjust the steam level will be repeated several times as the columns are loaded. Initially, steam flows should be set at a low level to prevent overloading the trays which might require shutdown and restart. During this period the valves in the reflux line are fully opened but the reflux pump is left off until enough liquid has built up in the condensate receiver. This prevents excessive wear on the pump. The reflux line *between* the two columns should also be opened and that reflux pump should be left off. The liquid level in the bottom of the beer still should be monitored and when it drops to half way, beer should be fed back into the column to refill the bottom of the still. The liquid level should continue to drop; if it does not, additional steam should be fed into the still bottom.

5. **Start reflux pump between the beer still and rectifying column.** When liquid starts to accumulate in the bottom of the rectifying column, the reflux pump between this still and the beer still is started. Flow in this line should be slow at first and then increase as more and more material reaches the rectifying column. When reflux is started to the beer still, the steam feed-rate will have to be slightly increased, because reflux tends to cool down a column.

6. **Start pump for reflux from the condensate receiver to rectifying column.** Eventually, enough vapor will have been condensed to fill the condensate receiver. Then, the pump for the reflux to the rectifying column can be started. Flow for this reflux line should be slow at first and then increased as more and more material distills. It should be noted that temperatures in the columns will be increasing as this process takes place. When the top temperature of the rectifying column is no longer increasing, the liquid levels in the bottom of the two columns are changing, and the condensate receiver level is no longer changing. Then, the reflux flow rates are at their designed flow and the column has reached equilibrium.

7. **Set beer feed pump, stillage pump, and product take-off at their designed flow.** Initially, the beer feed entering the beer still will be cooler than normal; the heat exchanger has not heated up yet. For this reason the steam to the beer still will need to be slightly increased. The thermocouple at the feed point will indicate when the feed is being heated to its designed temperature. At this time, the steam rate can be slightly lowered. Some minor adjustments will probably be needed. It must be kept in mind that this is a large system, and it takes some time for all points to react to a change in still conditions. All adjustments should be made, and then a period of time should be allowed before any additional adjustments are made.

8. **Check product quality.** Product quality at this time should be checked to insure that ethanol concentration is at the designed level. If it is lower than anticipated, the reflux ratio should be increased slightly. An increase in reflux cools the columns and additional heat must be applied to compensate for this. Also, the product flow-rate will be slightly decreased; therefore, flow rate to the still should also be varied. The ethanol concentration in the stillage should be checked to ensure that it does not exceed design concentration significantly.

9. **Dry ethanol.** After the ethanol leaves the distillation column, it must be further dried by passing through the molecular sieve drying columns and then stored in the ethanol storage tank. Literature from the vendor of the molecular sieve material will indicate at what temperature that flow must be switched to the other unit.

10. **Regenerate spent sieve material.** Carbon dioxide (CO_2) is used to regenerate the molecular sieve material. The CO_2 is collected from the fermentation system and compressed-CO_2 storage tank. To regenerate the molecular sieve material, the lines for regeneration are opened. Next, the CO_2 line is opened to allow flow to the stack heat exchanger and then on to the sieve columns. A rotameter in the CO_2 line is set to control the CO_2 flow-rate to the desired level. The molecular sieve columns are heated to about 450° F during regeneration. After regeneration is complete, the column is cooled down by CO_2 which bypasses the stack heat exchanger.

This essentially covers all the steps involved in the start-up of the plant. It should again be emphasized that caution must be exercised when operating any system of this complexity. If proper care is taken, and changes to the system operation are thought out sufficiently, successful plant operation will be achieved.

Shutdown. The second period of operation which differs significantly from normal operation is that period when the plant is being shut down. Proper care must be taken during shutdown to ensure both minimal losses of product and ease of restarting the process.

As the fermenters are individually shut down, they should be cleaned well to inhibit any unwanted microbial growth. The initial rinse from the fermenters can be pumped to the beer storage tank. Subsequent rinses should be discarded. The processing of this rinse material through the stills can continue until the top temperature of the beer column reaches 200° F. At this time, the unit should be put on total reflux.

During this shutdown period, the product quality will have degraded slightly, but the molecular sieve column will remove any additional water in the ethanol product. The stillage from the distillation system can be sent to the stillage storage system until the stillage is essentially clean. At this point, the steam to the column should be shut off and the column should be allowed to cool. During cooling, the column should be vented to prevent system damage. The pressure inside the column will be reduced as it cools. The air which enters the column at this time can be purged with steam prior to the next period of operation. The molecular sieve drying columns can be regenerated if necessary. The boiler should be shut down. If the shutdown period is of any significant duration, the boiler should be drained. If the plant is to be shut down for a short term, the fermenters should not require any additional cleaning. After an extended shutdown period, it is advisable to clean the fermenters in a manner similar to that performed at the initial start-up.

Shutdown periods are the best time to perform preventive maintenance. The column trays can be cleaned, pump seals replaced, etc. The important thing to remember is that safety must not be overlooked at this time. Process lines should be opened carefully because, even after extended periods of shutdown, lines can still be pressurized. If it is necessary to enter tanks, they must be well vented. It is suggested that an air line be placed in the tanks and that they be purged with air for several hours before they are entered. Also, a tank should never be entered without another person stationed outside the tank in case an emergency situation arises.

Daily Operation

The day-by-day operation of the representative on-farm plant requires the attention of the operator for two periods of about two hours each every day.

Each morning, the operator begins by checking the condition of the plant. All systems are operative because the operator would have been alerted by the alarm if there had been a shutdown during the night. A quick check will confirm that the beer flow and reflux flows are near desired values. The temperature of the top plate of the rectifying column and the proof of the product before drying should be checked. Even if the proof is low, the final product should be dry because the dryer removes essentially all of the water, regardless of input proof. However, excessively low entering proof could eventually overload the regeneration system. If the proof before drying is low, reflux flow is adjusted to correct it.

Next the fermenter that has completed fermentation is checked. The concentration of ethanol is checked and compared to the value indicated by the sugar content at the beginning of fermentation. If the concentration is suitable, the contents of the fermenter are dumped into the beer well. The inside of the fermenter is washed briefly with a high-pressure water stream. Then the fermenter is filled with preheated water from the holding tank.

The operator next checks the condition of the boiler and bale burner. The bale burner is reloaded with two of the large, round bales of corn stover from the row outside of the building. A front-end loader is used for this.

The operator returns to the fermenter that is being filled. It is probably half filled at this time, and the flow of meal into the tank is begun from the overhead meal bin. The flow rate is continuously measured and indicated, and will cut off when the desired amount is reached. The agitator in the tank is started. The liquefying enzyme is added at this time. The operator checks the temperature. When the tank is nearly full, steam is admitted to bring the temperature up to cooking value. The operator checks the viscosity until it is clear that liquefaction is taking place.

The operator now prepares for the automatically controlled sequence of the remaining steps of cooling and fermenting. The microprocessor controls these steps, and it will be activated at this time. However, the operator must load the saccharifying enzyme into its container. The enzyme is dumped into the fermentation tank on signal from the microprocessor. The yeast is pumped into the fermenter from the yeast tub, also on signal from the microprocessor. After cooking is complete, the microprocessor initiates the flow of cold water into coils in the vessel which cools the mash to the temperature corresponding to saccharification. When the appropriate temperature is reached, the enzyme is introduced. After a predetermined time, the converted mash is cooled to fermentation temperature, again by circulating cold water through the coils. When fermentation temperature is reached, the yeast is pumped into the fermenter. All of these operations are controlled by the microprocessor and do not require the operator's presence.

Once the fermentation is initiated, the operator can check the condition of the distillation columns and turn his/her attention to the products. The driver of the truck which delivers the whole stillage to the dairies and feeding operations will have finished filling the tank truck. If it is time for the pick-up of the ethanol, the operator will be joined by a field agent of the BATF who supervises the denaturing operation and checks the recorded flows of the plant to ensure that the product in storage is all that has been produced since the last pick-up. The distributions driver would then load the truck and start back to the bulk station.

In the evening the operator repeats the same operation with the exception of grinding meal and delivering the product.

MAINTENANCE CHECKLIST

Table V-5 provides a general timetable for proper maintenance of a representative ethanol plant.

TABLE V-5. MAINTENANCE CHECKLIST

Bale Burner		**Steam Lines**	
Remove ash	daily	Blow condensate	daily
Lubricate fans	monthly		
Check fan belts	monthly	**Beer Preheater**	
		Clean both sides	weekly
Water Softener			
Regenerate and backwash	weekly	**Beer Column**	
Check effectiveness	yearly	Clean out	weekly
Boiler		**Sight Glasses**	
Blow flues and CO_2 heater	monthly	Clean out	weekly
Check tubes and remove scale	monthly		
		Flow Meters	
Roller Mill		Clean out	as needed
Check for roller damage	weekly		
Check driver belts	monthly	**Condenser**	
		Descale water side	monthly
Elevator Leg to Meal Bin			
Lubricate	monthly	**Stillage Tank**	
		Clean and sterilize	monthly
Yeast Tubs			
Change air filter	monthly	**Pumps**	
		Check seals and end play	weekly
Fermenters		Lubricate	per manufacturer
Sterilize	every 3rd week		
Wash down outside	weekly	**Hydraulic System and Motors**	
		Check for leaks	daily
Back Pressure Bubblers		Change filter	per manufacturer
Clean out	weekly	Top-up	as necessary
Beer Well			
Sterilize and wash down	weekly		

REFERENCES

1. Ladisch, Michael R.; Dyck, Karen. "Dehydration of Ethanol: New Approach Gives Positive Energy Balance." *Science.* Vol. 205 (no. 4409): August 31, 1979; pp. 898–900.

CHAPTER VI
Business Plan

CHAPTER VI
BUSINESS PLAN

Preliminary planning is a prerequisite for the success of any project. Development of an ethanol plant involves planning not only the production process but also the management form and financial base.

The first step is to determine the financial requirements and relate that to the individual situation. From this the optimal organizational form can be selected, and the financing options can be examined.

The case study included in this chapter is an example of how a business plan may be completed. Every situation is different, however, and this can serve only as an example. The decision and planning worksheets at the end of Chapter II can be used in conjunction with the information in this chapter as tools in the decision-making process. The worksheets assist in analyzing financial requirements, choosing an organizational form, and selecting potential financing sources.

ANALYSIS OF FINANCIAL REQUIREMENTS

Financial requirements are determined by delineating capital costs, equity requirements, and operating costs. These requirements are then compared to potential earnings. Capital requirements include the costs for:

- real estate,
- equipment,
- business formation,
- installing equipment, and
- cost of licenses.

Although additional real estate may not be necessary, transfer of real estate to the business entity may be a consideration. Some of the equipment required for ethanol production has other farm uses and need not be charged totally to the ethanol production costs, e.g., grain bins and tractors with front-end loaders.

Equity requirements are established by the financial lending institution if borrowed capital is used. Equity can be in the form of money in savings, stocks and bonds, equipment, real estate, etc. Operating costs include:

- labor,
- maintenance,
- taxes,
- mortgages,
- supplies (raw materials, additives, enzymes, yeast, water),
- delivery expenses,
- energy (electricity and fuels),
- insurance and interest on short-term and long-term financing, and
- bonding.

Potential earnings are determined by estimating the sales price of ethanol per gallon and then multiplying by the number of gallons that the facility can sell, as well as income that may be derived from the sale of coproducts (determined by the sales price times the quantity that can actually be sold). Careful planning of markets for coproducts can significantly affect the net income of an ethanol production plant. In the case study projected financial statement included in this chapter, note the difference in net income based on different coproduct prices. In addition to actual income derived from the sale of ethanol and the coproducts, any savings realized by using ethanol to replace other fuels for on-farm use can be added to earnings. This is also true for coproducts such as stillage that might replace purchased feed.

Once the financial requirements and potential earnings are determined, they can be related to the specific situation. Capital costs and equity requirements are related to the individual capability to obtain financing. Operation costs are compared to potential earnings to illustrate cash flow.

Once this information is acquired and analyzed in relation to the individual's specific situation, a decision can be made about the organizational form for the production business.

ORGANIZATIONAL FORM

The organizational basis is the legal and business framework for the ethanol production facility. Broadly speaking, there are three principal kinds of business structures: proprietorship, partnership, and corporation.

A proprietor is an individual who operates without partners or other associates and consequently has total control of the business. A proprietorship is the easiest type of organization to begin and end, and has the most flexibility in allocating funds. Business profits are taxed as personal income and the owner/proprietor is personally liable for all debts and taxes. The cost of formation is low, especially in this case, since licensing involves only the BATF permit to produce ethanol and local building permits.

A partnership is two or more persons contractually associated as joint principals in a business venture. This is the simplest type of business arrangement for two or more persons to begin and end and has good budgetary flexibility (although not as good as a proprietorship). The partners are taxed separately, with profits as personal income, and all partners are personally liable for debts and taxes. A partnership can be established by means of a contract between two or more individuals. Written contractual agreements are not legally necessary, and therefore oral agreements will suffice. In a general partnership, each partner is personally liable for all debts of the partnership, regardless of the amount of equity which each partner has contributed.

A corporation is the most formalized business structure. It operates under the laws of the state of incorporation; it has a legal life all its own; it has its scope, activity, and name restricted by a charter; it has its profits taxed separately from the earnings of the executives or managers, and makes only the company (not the owners and managers) liable for debts and taxes. A Board of Directors must be formed and the purposes of the organization must be laid out in a document called "The Articles of Incorporation." Initial taxes and certain filing fees must also be paid. Finally, in order to carry out the business for which the corporation was formed, various official meetings must be held. Since a corporation is far more complex in nature than either a proprietorship or a partnership, it is wise to have the benefit of legal counsel. A corporation has significant advantages as far as debts and taxes are concerned. Creditors can only claim payments to the extent of a corporation's assets; no shareholder can be forced to pay off creditors out of his or her pocket, even if the company's assets are unequal to the amount of the debt.

There are often differences in the ease with which a business may obtain start-up or operating capital. Sole proprietors stand or fall on their own merits and worth. When a large amount of funding is needed, it may be difficult for one person to have the collateral necessary to secure a loan or to attract investors. Partnerships have an advantage in that the pooled resources of all the partners are used to back up the request for a loan and, consequently, it is often easier to obtain a loan because each and all of the partners are liable for all debts. Corporations are usually in the best position to obtain both initial funding and operating capital as the business expands. New shares may be issued; the company's assets may be pledged to secure additional funding; and bonds may be issued, backed up by the assets of the corporation.

A nonprofit cooperative is a special form of corporation. Such a cooperative can serve as a type of tax shelter. While the cooperative benefits each of its members, they are not held liable, either individually or collectively, for taxes on the proceeds from the sale of their products.

After determining the organizational form, financing options can be explored.

FINANCING

The specific methods of financing ethanol production plants can be divided into three general classes: private financing, public grants and loans, and foundations.

Private Financing

Private financing may be obtained from banks, savings and loan associations, credit unions, finance corporations, venture capital corporations, corporation stock issues, and franchise arrangements.

Foundations

Foundations provide funding either through grants or through direct participation by gaining equity, usually in the form of stocks in the production company. Often the investment portfolios that are used to generate income for foundations are composed in part of stocks in enterprises they deem appropriate to support the mission of the foundation.

Public Financing

Grants or loans are available from several federal agencies. See Table E-1, Sources of Public Financing contacts, in Appendix E. Each of the agencies has operating procedures and regulations that define appropriate use of their funding. The availability of funds varies from year to year.

CASE STUDY

The following case study of the Johnson family demonstrates the process for determining the feasibility of a farm-sized fermentation ethanol plant by developing a business plan. It is a realistic example, but the specific factors are, of course, different for every situation. This process may be used by anyone considering ethanol plant development, but the numbers must be taken from one's own situation. Table VI-1 delineates the assumptions used in the case study.

TABLE VI-1. CASE STUDY ASSUMPTIONS

- Corn is the basic feedstock.

- 25-gal EtOH/hr production rate.

- Operate 24 hrs/day; 5 days/week; 50 weeks/yr.

- Feed whole stillage to own and neighbors' animals.

- Sell ethanol to jobber for $1.74/gal.

- Sell stillage for 3.9¢/gal.

- Corn price is $2.30/bu (on-farm, no delivery charge, no storage fees).

- Operating labor is 4 hrs/day at $10/hr.

- Corn stover cost is $20/ton.

- Equity is $69,000.

- Debt is $163,040; at 15% per annum; paid semi-annually.

- Loan period is 15 yrs for plant; 8 yrs for operating capital and tank truck.

- Miscellaneous expenses estimated at 12¢/gal EtOH produced.

- Electricity costs estimated at 2¢/gal EtOH produced.

- Enzymes estimated at 4¢/gal EtOH produced.

Background Information

The Johnson family operates a 1,280-acre corn farm which they have owned for 15 years. They feed 200 calves in their feedlot each year. The family consists of Dave, Sue, and three children: Ted, 24 years old and married, has been living and working at the farm for 2 years, and he and his wife have a strong commitment to farming; Sara is 22 years old, married, and teaches in a town about 250 miles away; Laura is 15 years old, goes to high school in town 25 miles away, and also works on the farm.

The Johnsons are concerned about the future cost and availability of fuel for their farm equipment. The Johnsons have known about using crops to produce ethanol for fuel for a long time, and recent publicity about it has rekindled their interest.

They have researched the issues and believe there are five good reasons for developing a plan to build a fermentation ethanol plant as an integral part of their farm operation:

- to create another market for their farm products,

- to produce a liquid fuel from a renewable resource,

- to gain some independence from traditional fuel sources and have an alternate fuel available,

- to gain cost and fuel savings by using the farm product on the farm rather than shipping it, and by obtaining feed supplements as a coproduct of the ethanol production process, and

- to increase profit potential by producing a finished product instead of a raw material.

They first analyzed the financial requirements in relation to their location, farm operations, and personal financial situation.

Analysis of Financial Requirements

The local trade center is a town of 5,000 people, 35 miles away. The county population is estimated at 20,000. Last year 7,000,000 gallons of gasoline were consumed in the county according to the state gasoline tax department. A survey of the Johnsons' energy consumption on the farm for the last year shows:

Gasoline = 13,457 gals

Diesel = 9,241 gals

LP Gas = 11,487 gals

They decided to locate the plant close to their feedlot operation for ease in using the stillage for their cattle. They expect the plant to operate 5 days a week, 24 hours a day, 250 days a year. It is designed to produce 25 gallons of anhydrous ethanol per hour or 150,000 gallons in one year, using 60,000 bushels of corn per year. In addition to ethanol, the plant will produce stillage and carbon dioxide as coproducts at the rate of 230 gallons per hour.

After researching the question of fuel for the plant, this family has decided to use agricultural residue as the fuel source. This residue will be purchased from the family farm. The cost of this fuel is figured at $20 per ton.

They will have to purchase one year's supply of fuel since it is produced seasonally in the area. The tonnage of residue per acre available is 6 tons per acre based on measurements from the past growing season.

The water source is vitally important. They will need about 400 gallons of water each hour. To meet this demand, a well was drilled and an adequate supply of water was found. The water was tested for its suitability for use in the boiler, and the test results were favorable.

The family has determined that they can operate both the plant and the farm without additional outside labor. The ethanol they produce will be picked up at the farm as a return load by the jobbers making deliveries in the rural area. The Johnsons will deliver stillage in a tank truck to neighbors within 5 miles.

At the present there are no plans to capture the carbon dioxide, since the capital cost of the equipment is too high to give a good return on their investment. There is no good local market for the carbon dioxide; but there are many uses for carbon dioxide, and selling it as a coproduct may prove to be profitable in other situations.

The family's plans are to market their products locally. They have contacted local jobbers who have given them letters of intent to purchase the annual production of ethanol. They plan to use the distillers grains in their own feedlot and to sell the rest to their neighbors. The neighbors have given them letters stating they would purchase the remainder of the distillers grains produced. These letters are important in order to accurately assess the market.

Organizational Form

The Johnsons chose to establish a closely held corporation for this business. Other possibilities they considered included partnerships, sole proprietorships, and profit and nonprofit corporations. If additional equity had been needed, a broader corporation or a partnership would have been selected. However, their financial status was sufficient to allow them to handle the investment themselves, as shown by their balance sheet which follows.

Dave and Sue Johnson
Balance Sheet as of January 1979

Assets:

Current Assets:

Cash	$15,000
Inventory	$70,000
Total current assets	$85,000
Equipment	$125,000*
Land and buildings	$512,000*
Total assets	$722,000

Liabilities and Capital:

Operating notes at local bank	$20,000
Equipment loans at local bank	$69,000
Federal Land Bank loan	$350,000
Total Liabilities	$439,000
Total Capital	$283,000
Total Liabilities and Capital	$722,000

*These assets shown at fair market value.

The Johnsons formed a corporation because it afforded the ability to protect themselves from product liability and gave them the option to give stock to all family members as an incentive compensation package. Also, the use of a corporation avoids an additional burden on their credit line at the bank for their farming operation since they were able to negotiate a loan with no personal guarantee of the corporate debt. In a partnership they would have had personal liability for the product, the debt, and the actions of the partners in the business. The record-keeping requirements of the corporation and the limited partnership were equal, and the former afforded greater security. Co-ops and nonprofit corporations were considered also, but these two options were discarded because of operating restrictions.

After formation of the corporation, they transferred (tax-free) half of a year's supply of corn (30,000 bushels) in exchange for stock in the corporation. They elected Subchapter S treatment upon incorporation and had the first year of operation reported on a short-period return. Generally, Subchapter S has many of the advantages of a partnership but not the liabilities. (Consult an accountant or lawyer for a detailed description of this.) They could pass through the investment credit which is proposed to be 20% of the capital cost, assuming that the Internal Revenue Service would authorize a fuel-grade ethanol plant to qualify for the additional investment credit for being a producer of renewable energy. After the first short-period return is filed, the stockholders can then elect not to be a Subchapter S corporation. This plan helps the cash flow as they would personally recover some tax dollars through the investment credit.

The corporation will lease from the family, on a long-term basis, 2 acres of land on which to locate the plant. They considered transferring this land to the corporation, but the land is pledged as security for the Federal Land Bank so it would be cumbersome to get the land cleared of debt. Also, the 2 acres would require a survey and legal description, thereby adding additional cost, and there are no local surveyors who could do this work.

The corporation will purchase corn and agricultural residue from the family farm and damaged corn from neighbors when there is a price advantage to do so. The family will also purchase the distillers' grains and ethanol used on the farm from the corporation as would any other customer. All transactions between the family farm and the corporation will use current prices that would be paid or received by third parties.

Conclusion

The initial visit with the bank was encouraging. The local banker was well acquainted with ethanol production through the publicity it had been receiving. The bank was receptive to the financing, saying they would consider it an equipment loan. The bank required a schedule of production, funds-flow projections, projected income statements, and projected balance sheets for the next 2 years. The bank was primarily concerned that these statements demonstrate how the plant could be paid for.

Before meeting with their accountant, the Johnsons prepared decision and planning worksheets as described in Chapter II. This work on their part saved them some accountant's fees and gave them an idea as to the feasibility of such a plant. The projected financial statements were then prepared with the assistance of their accountant for the bank's use.

The following projected financial statement is based on decisions made about the operations and management of the plant. It served the Johnsons as a tool in deciding whether or not the plant would be a good investment for them and also as a final presentation to the bank for loan approval. The assumptions used in preparing these financial statements are included with the financial statements and represent an integral part of the management plan.

After the financial projections were completed and the bank had reviewed them, there was one more area of concern. The bank wanted to know whether the system as designed was workable and could produce what it was projected to do from a technical feasibility standpoint. The family furnished the bank with the engineer's report which documented systems that were in operation and that were successfully using their proposed technology. The bank contacted some of the people operating these plants to verify their production. The bank then completed their paperwork and made a loan to the family's corporation secured only by the equipment. They also approved the line of credit for the working capital required based on the projected financial statements.

GALUSHA HIGGINS & GALUSHA

GLASGOW, MONTANA

November 6, 1979

National Bank of Golden Rise
Golden, Colorado

We have assisted in the preparation of the accompanying projected balance sheet of Johnson Processors, Inc. (a sample company), as of December 31, 1980 and 1981, and the related projected statements of income and changes in financial position for the years then ended. The projected statements are based solely on management's assumptions and estimates as described in the footnotes.

Our assistance did not include procedures that would allow us to develop a conclusion concerning the reasonableness of the assumptions used as a basis for the projected financial statements. Accordingly, we make no representation as to the reasonableness of the assumptions.

Since the projected statements are based on assumptions about circumstances and events that have not yet taken place, they are subject to the variations that will arise as future operations actually occur. Accordingly, we make no representation as to the achievability of the projected statements referred to above.

The terms of our engagement are such that we have no obligation or intention to revise this report or the projected statements because of events and transactions occurring after the date of the report unless we are subsequently engaged to do so.

G, H & G
Certified Public Accountants

JOHNSON PROCESSORS, INC.
PROJECTED BALANCE SHEET
UNAUDITED

(NOTES 1 THROUGH 4 ARE AN INTEGRAL PART OF THESE PROJECTED FINANCIAL STATEMENTS AND PROVIDE AN EXPLANATION OF ASSUMPTIONS USED IN THIS REPORT)

		FOR YEAR ENDED		
		December 31, 1980		December 31, 1981
Assets				
Current assets				
Cash		$24,617		$18,452
Accounts receivable		55,350		55,350
Raw materials and supplies		22,214		26,785
Work in process inventory		1,362		1,601
Finished goods inventory		18,765		22,041
Marketable securities		30,000		30,000
Total current assets		$152,308		$154,229
Plant, equipment, and structures				
Plant and equipment	$107,000		$107,000	
Building	17,280		17,280	
Total plant and equipment	$124,280		$124,280	
Less accumulated depreciation	8,605		17,210	
Net plant and equipment		115,675		107,070
Total assets		$267,983		$261,299
Liabilities and Capital				
Current liabilities				
Accounts payable		$3,800		$4,130
Current portion of loans		2,997		3,500
Total current liabilities		$6,797		$7,630
Long-term liabilities				
Bank loan	$156,043		$113,821	
Less current portion	2,997		3,500	
Total long-term liabilities		153,046		110,321
Total liabilities		$159,843		$117,951
Capital		108,140		143,348
Total liabilities and capital		$267,983		$261,299

JOHNSON PROCESSORS, INC.
PROJECTED INCOME STATEMENT
UNAUDITED

(NOTES 1 THROUGH 4 ARE AN INTEGRAL PART OF THESE PROJECTED FINANCIAL STATEMENTS AND PROVIDE AN EXPLANATION OF ASSUMPTIONS USED IN THIS REPORT)

	FOR YEAR ENDED	
	December 31, 1980	December 31, 1981
Revenue		
Alcohol	$229,680	283,500
Stillage	55,525	55,973
Total sales	$285,205	$339,473
Cost of goods sold		
Beginning finished goods inventory	0	$18,765
Cost of goods manufactured	$225,634	266,634
Cost of goods available for sale	$225,634	$285,399
Ending finished goods inventory	18,765	22,041
Cost of goods sold	$206,869	$263,358
Gross profit	$78,336	$76,115
Selling expenses		
Marketing and delivery expenses (scheduled)	$25,545	$29,074
Total selling expenses	$25,545	$29,074
Net operating profit	$52,791	$47,041
Income taxes	13,651	11,833
Net income	$39,140	$35,208

JOHNSON PROCESSORS, INC.
PROJECTED STATEMENT OF CHANGES IN FINANCIAL POSITION (Cash Basis)
UNAUDITED

(NOTES 1 THROUGH 4 ARE AN INTEGRAL PART OF THESE PROJECTED FINANCIAL STATEMENTS AND PROVIDE AN EXPLANATION OF ASSUMPTIONS USED IN THIS REPORT)

	FOR YEAR ENDED	
	December 31, 1980	December 31, 1981
CASH GENERATED		
Net income	$39,140	$35,208
Add (deduct) items not requiring or generating cash during the period		
Trade receivable increase	(55,350)	(0)
Trade payable increase	3,800	330
Inventory increase	(42,341)	(8,086)
Depreciation	8,605	8,605
Subtotal	$(46,146)	$36,057
Other sources		
Contributed by shareholders	69,000	
Bank loan	163,040	
Total cash generated	$185,894	$36,057
CASH APPLIED		
Additional loan repayment	$ 4,000	$38,722
Purchase of plant and equipment	107,000	
Purchase of building	17,280	
Reduction of bank loan	2,997	$3,500
Total cash applied	$131,277	$42,222
Increase in cash	$ 54,617	$(6,165)*

*Net decrease in cash is caused by an accelerated pay-off of the operating capital rate in the amount of $38,722.

JOHNSON PROCESSORS, INC.
EXHIBIT I
UNAUDITED

(NOTES 1 THROUGH 4 ARE AN INTEGRAL PART OF THESE PROJECTED FINANCIAL STATEMENTS AND PROVIDE AN EXPLANATION OF ASSUMPTIONS USED IN THIS REPORT)

	FOR YEAR ENDED	
	December 31, 1980	December 31, 1981

ETHANOL

Projected Production Schedule

Projected gallons sold	132,000	150,000
Projected inventory requirements	18,000	18,000
Total gallons needed	150,000	168,000
Less inventory on hand	0	18,000
Projected production	150,000	150,000
Sales price per gallon	$1.74	$1.89

Projected Cost of Goods Manufactured

Projected production costs:		
Labor	$20,805	$20,328
Corn	138,000	180,000
Electricity	3,000	3,300
Straw	10,714	11,785
Miscellaneous (scheduled)	18,000	19,800
Depreciation	6,730	6,730
Interest	23,747	18,330
Enzymes	6,000	6,600
Total costs of production	$226,996	$266,873
Add beginning work-in-process inventory		1,362
Subtotal	$226,996	$268,235
Less ending work-in-process inventory	1,362	1,601
Projected cost of goods manufactured	$225,634	$266,634

JOHNSON PROCESSORS, INC.
NOTES TO THE PROJECTED FINANCIAL STATEMENTS
UNAUDITED

1. SIGNIFICANT ACCOUNTING POLICIES

Following is a summary of the significant accounting policies used by Johnson Processors, Inc. in the projected financial statements.

- Assets and liabilities, and revenues and expenses are recognized on the accrual basis of accounting.

- Inventory is recorded at the lower value (cost or market) on the first-in, first-out (FIFO) basis.

- Accounts receivable are recorded net of bad debts.

- Depreciation is calculated on the straight line basis.

2. ASSETS

Current Assets

- Accounts receivable are projected at each balance sheet date using 30 days of sales for ethanol and 90 days of sales for stillage.

- Inventory—raw materials—is made up of corn and corn stover. Thirty days in inventory is used for corn and one year's supply is used for stover.

- Inventory—work in process—consists of 1½ days' production.

- Inventory of finished goods consists of raw materials and cost of production. Thirty days in inventory is used for ethanol and 2 days is used for stillage.

The estimates of number of days in accounts receivable and finished goods inventory are higher than those quoted in Robert Morris Associates averages for feed manufacturers and wholesale petroleum distributors.

Fixed Assets

Management anticipates purchasing the equipment for production of ethanol. Consulting engineers contacted verified that the equipment and plant costs listed in Table VI-2 were reasonable.

REPRESENTATIVE PLANT COSTS

Equipment and Materials	$ 71,730
Piping	4,000
Electrical	1,500
Excavation and Concrete	2,000
Total Equipment and Materials	79,230
10% Contingency	7,923
Total	87,153
Tank Truck	14,847
Erection Costs	5,000
Grand Total	$107,000

Investments

Investments consist of the amount of excess cash accumulated from operation during the first and second year of operation.

3. LIABILITIES AND CAPITAL

Management estimated accounts payable using 30 days in payables for conversion costs. It is anticipated that corn will be paid for a month in advance.

Management estimates that a bank loan in the amount of $163,040 will be required, payable semiannually at 15% interest. Anticipated payback period for the portion of the loan covering plant and equipment is 15 years. The payback period for the portion covering working capital is 8 years. The payback period for the truck is 8 years. An additional $4,000 the first year is projected to be paid on the equipment loans and to repay the working capital loan in the second year. The loan will be used to finance plant and equipment and working capital. The anticipated plant and equipment and working capital for the first year is estimated as follows.

Plant and equipment	$124,280
Working capital	$107,760
Total	$232,040

Cash could be very lean during the first year that the plant operates at capacity because of dramatic increases in working capital resulting from accounts receivable and inventory requirements. Inadequate financing would make maximum production impossible because of inability to fund working capital demands.

4. INCOME STATEMENT
Sales

Sales volume was estimated at maximum production (150,000 gallons of ethanol and 1,380,000 gallons stillage) for the first year. Ethanol price was taken to be $1.74 per gallon (the actual delivered price at Council Bluffs, Iowa on November 6, 1979). The price of ethanol is projected to increase by 9.1% for the entire period covered by the projections. The increase of 9.1% is the projected price increase by a marketing firm from Louisiana.

It is conceivable that as the price of gasoline increases to a point greater than the price of ethanol, producers could raise the price of ethanol to equalize the prices of the two liquid fuels. In order to be conservative, management did not consider this effect .

The stillage sales price was taken to be 3.9 cents per gallon for the 2 years. This sales price was based on the sales price charged by Dan Weber in Weyburn, Saskatchewan, Canada. The local stillage market is, however, worthy of a thorough study before a decision is made to enter the fermentation ethanol business. If a large brewery or distillery is located in the area, the price of stillage can be severely depressed. For example, Jack Daniels and Coors sell their stillage for 0.4 cents per gallon and 0.8 cents per gallon, respectively.

Cost of Sales

Management has projected cost of sales to include raw materials and production costs. The cost of corn is projected at $2.30 per bushel during the first year of operation (the price received by farmers in Iowa on November 6, 1979). Management has the total amount of corn available from the corporate shareholder. To demonstrate the effect of substantial increases in corn prices on profitability and cash flow, management projected that the cost of corn would rise to $3.00 per bushel for year two.

Management anticipates that depreciation will remain constant using the straight line method. A 15-year life for the plant was used with a salvage value of $4,000, while an 8-year life and 20-year life were used for the truck and building respectively. No other salvage values were taken into consideration.

Labor cost was computed allowing 4 hours per day for work necessary in the processing of the ethanol, based on the engineer's time requirement estimates. The labor was valued at $10 per hour, including a labor overhead factor. Bookkeeping labor was computed at $6,000 per year, assuming this plant would only require part-time services. It is anticipated that some additional time may be required the first year. For this, $2,325 has been added to the labor cost as a contingency.

Enzyme cost was estimated at $6,000 per year by the engineers working on the project. Electricity was estimated at 2 cents per gallon of ethanol produced.

Stover cost was computed based on a cost of $20 per ton. A Btu value of 7,000 Btu per pound (as estimated by the engineers) was used. An 80% efficiency for the boiler was assumed, so anticipated Btu values were 5,600 Btu per pound of straw, and a total Btu requirement of 40,000 per gallon of ethanol produced.

Miscellaneous Expenses

Miscellaneous expenses were estimated at 12 cents per gallon. These expenses are shown in the detail schedule at the end of this report. To be conservative, figures are included in the miscellaneous expenses for shrinkage due to the grain handling and a contingency for any minor items that may have been overlooked.

Interest expense is for the bank loan. Interest expense is calculated at 15%. In year two of the operation, it is projected that the working capital portion of the notes payable will be paid off.

Management projects that other projected costs will increase 10% per year because of inflation. To be conservative, management did not estimate the cost savings potential of improved technology. Research is currently being performed in crops that have the potential of producing several times the amount of ethanol as does corn. Use of such crops could produce substantial cost savings in ethanol production. The process is very new in design, so improvements in the production process are also probable. Such improvements could further reduce the cost of producing ethanol.

Selling and Administrative Expenses

To be conservative, management estimated marketing expenses at 5% of sales. It is anticipated that this expense may not actually be necessary. Delivery expenses take into consideration the following items:

- interest was computed at 15% on the bank loan for the truck based on semi-annual payments;

- the time necessary to deliver the stillage was estimated based on 10 hours per week at $10 per hour including labor overhead;

- maintenance for the truck was estimated at $1,000; and

- fuel for the truck was computed based on 75 miles per week and a fuel consumption of 4 miles per gallon and a fuel cost of $1.025 per gallon. The cost is estimated to increase 36.5% for the second year of operation.

Income Taxes

The shareholders of Johnson Processors, Inc. plan to elect to have income taxed to the shareholder rather than to the corporation, under Internal Revenue Code Section 1372(a). The shareholders anticipate changing the election after the first year of operation. Taxes have been estimated based on a 6% state tax rate and a 6.75% federal tax rate in effect during 1979. For purposes of these projections, the projected financial statements (assuming a conventional corporation and a full 12 months of operation in each period) have been shown to demonstrate the projected results of operation that could be anticipated.

The shareholders of Johnson Processors, Inc. anticipate contributing $69,000 to the corporation. This amount is 30% of the total project. This will be contributed by transferring corn inventory equal to the ½-year supply necessary for processing. For purposes of this illustration, the contribution of corn is treated as cash to demonstrate to the bank the payback potential of the plant.

JOHNSON PROCESSORS, INC.
DETAIL SCHEDULES
UNAUDITED

(NOTES 1 THROUGH 4 ARE AN INTEGRAL PART OF THESE PROJECTED FINANCIAL STATEMENTS AND PROVIDE AN EXPLANATION OF ASSUMPTIONS USED IN THIS REPORT)

	FOR YEAR ENDED	
	December 31, 1980	December 31, 1981
Schedule of Marketing and Delivery Expenses		
Marketing 5% of sales	$14,260	$16,974
Interest on truck	2,211	2,045
Depreciation	1,875	1,875
Labor	5,200	5,720
Maintenance	1,000	1,100
Fuel	1,000	1,360
Total	$25,546	$29,074
Schedule of Miscellaneous Expenses		
Property taxes	$2,250	$2,475
Insurance	2,100	2,310
Chemicals and supplies	600	660
Yeast	450	495
Shrinkage	4,200	4,620
Other	2,550	2,805
Contingencies	5,850	6,435
Total	$18,000	$19,800

Appendices

APPENDIX A

Environmental Considerations

- **Areas for Potentially Hazardous Environmental Effects**
  - **Ethanol Production**
  - **Vehicular Fuel Use**

AREAS FOR POTENTIALLY HAZARDOUS ENVIRONMENTAL EFFECTS

There are two areas to be examined for potentially hazardous environmental effects—the ethanol production process and large-scale use of ethanol as a vehicular fuel.

Ethanol Production

There will be several types of effects on the environment caused by the production of fermentation ethanol which can be avoided with proper precautions.

Crop Residue Removal. The first and possibly most important environmental impact could be the removal of crop residues for use as a boiler fuel. Crop residues are important because they help control soil erosion through their cover and provide nutrients, minerals, and fibrous material which help maintain soil quality. However, not more than one-third to one-half of the residues from a grain crop devoted to ethanol production need be used to fuel the process. Also, there are several methods, such as crop rotation and winter cover crops, which lessen the impact of crop residue removal.

Use of Wet Stillage. The second environmental impact to be considered is related to the application of thin stillage to the land. Thin stillage, the product of the filtering process whereby the course solids are filtered out of the whole stillage, is composed of very small solid particles and solubles. Two kinds of problems can result from applying thin stillage to the land: odor and acidity. The impacts of applying thin stillage to the land can be attenuated by using a sludge plow, possible recycling of the thin stillage within the plant, or use of anaerobic digestion to reduce the pollution potential of the thin stillage.

Air Pollution. The third potential environmental impact is air pollution. Two forms of air pollution could result from development of an ethanol production scheme on-farm: the release of pollutants from the boiler used to produce steam from process heat, and vaporization of ethanol lost during the production process. If crop residues are used as boiler fuel, which is a preferred plan, the resulting pollutants are primarily particulate matter which can be controlled through the use of flue stack scrubbers. The release of ethanol vapors is not a major concern at this time.

Vehicular Fuel Use

Under the Clean Air Act of 1977 all fuel additives are automatically banned unless the manufacturer of the fuel additive demonstrates that the additive will not cause or contribute to the failure of any vehicle to meet applicable emission standards. The federally specified reference fuel is indolene and the required tests are for emissions of tailpipe hydrocarbons, carbon monoxide, nitrogen oxides, and evaporative hydrocarbons. The last mentioned test involves the measurement of fuel vapors from the gas tank when the engine is left to idle and from locations such as the pump at the gas station. The requirements in these tests are quite stringent; in fact, in an Environmental Protection Agency (EPA) study [1], commonly accepted pure summer-grade gasoline when compared with indolene failed—in oxidation catalyst vehicles and in three-way catalyst vehicles—all four tests mentioned above.

On the other hand, the same study reported that a 90%-indolene/10%-ethanol blend showed definite improvements over pure indolene in tailpipe hydrocarbon and carbon monoxide emissions, nitrogen oxides emissions equal to the summer-grade gasoline, and a slight increase of evaporative emissions relative to the pure summer-grade gasoline.

EPA and the Department of Energy have conducted a cooperative gasohol testing program to obtain and evaluate environmental impact data. On the basis of these tests, EPA concluded that the addition of 10% ethanol to gasoline [2]:

- slightly decreases hydrocarbon emissions;

- significantly decreases carbon monoxide emissions;

- slightly increases nitrogen oxides emissions; and

- substantially increases evaporative hydrocarbon emissions.

On December 16, 1978, EPA approved use of gasohol under Section 211(f)(3) of the Clean Air Act of 1977 and found that there was no significant environmental risk associated with gasohol's continued use [3]. Furthermore, new emissions control systems, such as the "threeway catalyst with exhaust oxygen sensors for carburetion feedback for air-fuel control," have been shown to be equally effective using either gasoline or gasohol [3].

In summary, the results to date have been generally favorable with respect to the use of gasohol in automobiles. However, in a recent technical memorandum, the Office of Technology Assessment of the Congress of the United States stated that the "mixture of observed emissions reductions and increases, and the lack of extensive and controlled emissions testing, does not justify a strong value judgment about the environmental effect of gasohol used in the general automobile population (although the majority of analysts have concluded that the net effect is unlikely to be significant)" [4].

REFERENCES

1. Jackson, B. R. "Testimony at DOE Hearings on Alcohol Fuels." U.S. Environmental Protection Agency; December 6, 1978.

2. *Characterization Report: Analysis of Gasohol Fleet Data to Characterize the Impact of Gasohol on Tailpipe and Evaporative Emissions.* U.S. Environmental Protection Agency, Technical Support Branch, Mobile Source Enforcement Division, December 1978.

3. Allsup, J. R.; Eccleston, P. B. "Ethanol/Gasoline Blends as Automotive Fuels." Alcohol Fuels Technology, 3rd International Symposium; Asilomar, CA; 1979.

4. Office of Technology Assessment. *Gasohol—A Technical Memorandum.* September 1979, 69 p. Available from Superintendent of Documents, U.S. Government Printing Office, Stock No. 052-003-00706-1. [P48]

Reference Information

- **Conversion Factors**

- **Properties of Ethanol, Gasoline, and Water**

- **Water, Protein, and Carbohydrate Content of Selected Farm Products**

- **Typical Analysis of Distillers' Dried Grain Solids–Corn**

- **Energy Content for Feed**

- **Percent Sugar and Starch in Grains**

- **Comparison of Raw Materials for Ethanol Production**

- **Comparative Energy Balances for Ethanol Production**

CONVERSION FACTORS

1 gallon water = 8.33 pounds (at 60 °F) = 0.134 cubic foot = 128 fluid ounces = 4 quarts = 8 pints = 3.785 liters

1 gallon of ethanol weighs 6.6 pounds

1 barrel of crude oil = 42 gallons

1 Btu = 252 calories = heat required to raise 1 lb of water 1 degree Fahrenheit (°F).

1 bushel = 56 pounds

1 calorie = .00397 Btu = heat required to raise 1 gram of water 1 degree Centigrade (°C)

1 U.S. liquid gallon = 4 quarts = 231 cubic inches = 3.78 liters

8 gallons = 1 bushel

1 liter = 1.057 U.S. liquid quarts

1 fluid ounce = 30 milliliters

1 pound = 453.6 grams

1 cubic foot = 7.48 liquid gallons = 62.36 pounds H_2O (at 60° F)

1 acre = 43,560 square feet = 4,840 square yards

To convert from °F to °C, subtract 32 and then multiply by ⅝.

To convert from °C to °F, multiply by ⅝ and then add 32.

PROPERTIES OF ETHANOL, GASOLINE AND WATER

CHEMICAL PROPERTIES	Gasoline	Ethanol	Water
Formula	C_4-C_{12}	CH_3CH_2OH	H_2O
Molecular Weight	varies	46.1	18.015
%Carbon (by weight)	85–88	52.1	
%Hydrogen (by weight)	12–15	13.1	
%Oxygen (by weight)	indefinite	4.7	
C/H ratio	5.6–7.4	4.0	
Stoichiometric Air-to-Fuel Ratio	14.2–15.1	9.0	

PHYSICAL PROPERTIES	Gasoline	Ethanol	Water
Specific Gravity	0.70–0.78	0.794	1.0
Liquid Density lb/ft³	43.6 approx.	49.3	
lb/gal	5.8-6.5	6.59	
psi at 100° F (Reid)	7–15	2.5	
psi at 77° F	0.3 approx.	0.85	
Boiling Point (°F)	80–440	173	212
Freezing Point (°F)	−70 approx.	−173	32
Solubility in Water	240 ppm	infinite	
Water in	88 ppm	infinite	
Surface Tension (dyne/cm²)		23	54.9
Dielectric Constant		24.3	
Viscosity at 68°F (cp)	0.288	1.17	1.0
Specific Resistivity	2×10^{16}	0.3×10^6	

THERMAL PROPERTIES	Gasoline	Ethanol	Water
Lower Heating Value			
Btu/lb	18,900 (avg)	11,500	
Btu/gal	115,400 (avg)	73,560	
Higher Heating Value			
Btu/lb at 68°F	20,260	12,800	
Btu/gal	124,800	84,400	
Heat of Vaporization			
Btu/lb	150	396	940
Btu/gal	900	3,378	7,802
Octane Ratings			
Research	91–105	106–108	
Pump (Ron + Mon)/2	86–90	98–100	
Flammability Limits			
(% by volume in air)	1.4–7.6	4.3–19.0	
Specific Heat			
(Btu/lb − °F)	0.48	0.60	1.0
Autoignition Temperature (°F)	430–500	685	
Flash Point (°F)	−50	70	
Coefficient of Thermal			
Expansion at 60°F and 1 atm	0.0006	0.00112	0.00

WATER, PROTEIN, AND CARBOHYDRATE
CONTENT OF SELECTED FARM PRODUCTS

Crop	%Water	%Protein	% Carbo-hydrate	Crop	%Water	%Protein	% Carbo-hydrate
Apples, raw	84.4	0.2	14.5	Muskmelons	91.2	0.7	7.5
Apricots, raw	85.3	1.0	12.8	Mustard greens	89.5	3.0	5.6
Artichokes, French	85.5	2.9	10.6	Okra	88.9	2.4	7.6
Artichokes,				Onions, dry	89.1	1.5	8.7
Jerusalem	79.8	2.3	16.7	Oranges	86.0	1.0	12.2
Asparagus, raw	91.7	2.5	5.0	Parsnips	79.1	1.7	17.5
Beans, lima, dry	10.3	20.4	64.0	Peaches	89.1	0.6	9.7
Beans, white	10.9	22.3	61.3	Peanuts	5.6	26.0	18.6
Beans, red	10.4	22.5	61.9	Pears	83.2	0.7	15.3
Beans, pinto	8.3	22.9	63.7	Peas, edible pod	83.3	3.4	12.0
Beets, red	87.3	1.6	9.9	Peas, split	9.3	1.0	62.7
Beet greens	90.9	2.2	4.6	Peppers, hot chili	74.3	3.7	18.1
Blackberries	84.5	1.2	12.9	Peppers, sweet	93.4	1.2	4.8
Blueberries	83.2	0.7	15.3	Persimmons	78.6	0.7	19.7
Boysenberries	86.8	1.2	11.4	Plums, Damson	81.1	0.5	17.8
Broccoli	89.1	3.6	5.9	Poke shoots	91.6	2.6	3.1
Brussels sprouts	85.2	4.9	8.3	Popcorn	9.8	11.9	72.1
Buckwheat	11.0	11.7	72.9	Potatoes, raw	79.8	2.1	17.1
Cabbage	92.4	1.3	5.4	Pumpkin	91.6	1.0	6.5
Carrots	8.2	1.1	9.7	Quinces	83.8	0.4	15.3
Cauliflower	91.0	2.7	5.2	Radishes	94.5	1.0	3.6
Celery	94.1	0.9	3.9	Raspberries	84.2	1.2	13.6
Cherries, sour	83.7	1.2	14.3	Rhubarb	94.8	0.6	3.7
Cherries, sweet	80.4	1.3	17.4	Rice, brown	12.0	7.5	77.4
Collards	85.3	4.8	7.5	Rice, white	12.0	6.7	80.4
Corn, field	13.8	8.9	72.2	Rutabagas	87.0	1.1	11.0
Corn, sweet	72.7	3.5	22.1	Rye	11.0	12.1	73.4
Cowpeas	10.5	22.8	61.7	Salsify	77.6	2.9	18.0
Cowpeas, undried	66.8	9.0	21.8	Soybeans, dry	10.0	34.1	33.5
Crabapples	81.1	0.4	17.8	Spinach	90.7	3.2	4.3
Cranberries	87.9	0.4	10.8	Squash, summer	94.0	1.1	4.2
Cucumbers	95.1	0.9	3.4	Squash, winter	85.1	1.4	12.4
Dandelion greens	85.6	2.7	9.2	Strawberries	89.9	0.7	8.4
Dates	22.5	2.2	72.9	Sweet potatoes	70.6	1.7	26.3
Dock, sheep sorrel	90.9	2.1	5.6	Tomatoes	93.5	1.1	4.7
Figs	77.5	1.2	20.3	Turnips	91.5	1.0	6.6
Garlic cloves	61.3	6.2	30.8	Turnip greens	90.3	3.0	5.0
Grapefruit pulp	88.4	0.5	10.6	Watermelon	92.6	0.5	6.4
Grapes, American	81.6	1.3	15.7	Wheat, HRS	13.0	14.0	69.1
Lamb's-quarters	84.3	4.2	7.3	Wheat, HRW	12.5	12.3	71.7
Lemons, whole	87.4	1.2	10.7	Wheat, SRW	14.0	10.2	72.1
Lentils	11.1	24.7	60.1	Wheat, white	11.5	9.4	75.4
Milk, cow	87.4	3.5	4.9	Wheat, durum	13.0	12.7	70.1
Milk, goat	87.5	3.2	4.6	Whey	93.1	0.9	5.1
Millet	11.8	9.9	72.9	Yams	73.5	2.1	23.2

Source: *Handbook of the Nutritional Contents of Foods,* USDA.

TYPICAL ANALYSIS OF DISTILLERS' DRIED GRAIN SOLIDS—CORN

	Distillers' Dried Grains (%)	Distillers' Dried Solubles (%)	Distillers' Dried Grains with Solubles (%)
Moisture	7.5	4.5	9.0
Protein	27.0	28.5	27.0
Lot	7.6	9.0	8.0
Fiber	12.8	4.0	8.5
Ash	2.0	7.0	4.5
Amino Acids			
Lysine	0.6	0.95	0.6
Methionine	0.5	0.5	0.6
Cystine	0.2	0.4	0.4
Histidine	0.6	0.63	0.6
Arginine	1.1	1.15	1.0
Aspartic Acid	1.68	1.9	1.7
Threonine	0.9	0.98	0.95
Serinine	1.0	1.25	1.0
Glutamic Acid	4.0	6.0	4.2
Proline	2.6	2.9	2.8
Glycine	1.0	1.2	1.0
Alanine	2.0	1.75	1.9
Valine	1.3	1.39	1.3
Isoleucine	1.0	1.25	1.0
Leucine	3.0	2.6	2.7
Tyrosine	0.8	0.95	0.8
Phenylalanine	1.2	1.3	1.2
Tryptophan	0.2	0.3	0.2

ENERGY CONTENT FOR FEED

	Distillers' Dried Grains	Distillers' Dried Solubles	Distillers' Dried Grains with Solubles
For Cattle:			
Total digestible nutrients	83	80	82
Megacalories per kilogram	2.19	2.32	2.3
For Poultry:			
Megacalories per kilogram	2.0	2.75	2.62
For Swine:			
Megacalories per kilogram	1.84	2.98	3.39

Source: *Feed Formulation*, Distillers Feed Research Council, 1435 Enquirer Building, Cincinnati, Ohio 45202.

PERCENT SUGAR AND STARCH IN GRAINS

Grain	%Sugar	%Starch
Barley	2.5	64.6
Corn	1.8	72.0
Grain sorghum	1.4	70.2
Oats	1.6	44.5
Rye	4.5	64.0
Wheat		63.8

Source: *Composition of Cereal Grains and Forages,*
National Academy of Sciences publication.

COMPARISON OF RAW MATERIALS FOR ETHANOL PRODUCTION

Raw Material	Gal Ethanol	Protein Yield	%Protein dry
Corn	2.6/bu	18lb/bu	29–30
Wheat	2.6/bu	20.7/bu	36
Grain sorghum	2.6/bu	16.8/bu	29–30
Average starch grains	2.5/bu	17.5/bu	27.5
Potatoes(75% Moist)	1.4/cwt	14.8/cwt	10
12–14			
Sugar beets	20.3/ton	264/ton	20
Molasses (50% sugar)	0.4/gal	68/ton	20

Source: National Gasohol Commission.

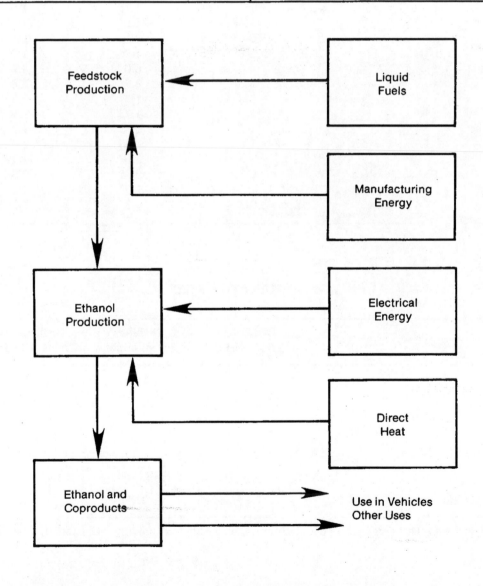

Figure D-1. Ethanol Production System Block Diagram

COMPARATIVE ENERGY BALANCES FOR ETHANOL PRODUCTION

Energy balances are a confusing and controversial subject. The sources of the confusion are varied but most stem from differences in opinion regarding what must be included for consideration and the proper approach to use. One of the principal sources of confusion is the type of energy balance being investigated in a given case. Some energy balance studies compare the total energy contents of the products and coproducts with the fossil energy consumed in their production. Other studies compare the amount of crude petroleum energy required to produce a given amount of petroleum substitute. Whatever types of energy are compared, an energy balance study has the objective to compare the energy input to a system with the energy output of the system. If the energy input is greater than the energy output, the energy balance is said to be negative; conversly, if the energy output of the system is greater than the energy input to the system, the energy balance is said to be positive. The causes of disparity among the various studies include differences in assumptions and reference technologies, and ambiguities in defining the boundaries of the given system under consideration.

Consider the ethanol production system shown in Figure D-1. Energy inputs to the system include the liquid fuel and manufacturing energy required to produce the feedstocks and the electrical and heat energy required to convert the feedstocks into ethanol. Note that the solar energy input is not included. Energy output of the system is in the form of ethanol which can be used in vehicles and other applications and other coproducts. To illustrate how differences in opinion among various studies can arise, consider the ethanol energy balance studies of Scheller and Mohr [1] and Reilly [2]. For 1 bushel of corn, the two studies calculate similar values for the total nonrenewable energy inputs as follows:

ENERGY BALANCES
(Basis: 1 Bushel Corn)

Energy Inputs	Scheller & Mohr	Reilly
Agricultural Energy	119,000 Btu	135,000 Btu
Direct on-farm		
Fertilizer and chemicals		
Transport		
Ethanol Process Energy	370,000 Btu	368,000 Btu
Cooking and fermentation	64,000	
Distilling and centrifuging	105,000	
Dehydration	37,000	
Evaporation of stillage	113,000	
Drying of stillage	51,000	
TOTAL ENERGY INPUT	489,000 Btu	503,000 Btu

From similar values of energy input, Scheller and Mohr proceed to calculate a positive energy balance, while Reilly calculates a negative energy balance. Reilly considers the outputs to be 2.6 gallons of ethanol, with a total (lower) heating value of about 191,000 Btu, and the stillage coproduct which can be given an energy credit of about 49,000 Btu [3]. Subtracting the energy input of 503,000 Btu from the total of energy output of 240,000 Btu, Reilly obtains a negative energy balance of 260,000 Btu. However, Scheller and Mohr include, as an additional coproduct, the heat content of 75% of the corn stover to be used as energy input into the ethanol production process. This amounts to an additional energy output of about 322,000 Btu. Thus, Scheller and Mohr would calculate a total energy output of 562,000 Btu and achieve a positive energy balance of about 73,000 Btu for each bushel of corn processed into ethanol.

REFERENCES

1. Scheller, W., Bohr B. "Net Energy Balance for Ethanol Production." Presented at the 171st National Meeting of the American Chemical Society, New York, April 7, 1976.

2. Reilly, P.J. "Economics and Energy Requirements for Ethanol Production." Department of Chemical and Nuclear Engineering, Iowa State University; January 1978.

3. Chambers, R.S., Herendeen, R.A., Joyce, J.J., and Penner, P.S. "Gasohol: Does It or Doesn't It Produce Positive Net Energy?" *Science*, Volume 206 (no. 4420): November 16, 1979, pp. 789–795.

APPENDIX C

Bibliography

- **General**

- **Conversion**
- **Coproducts**
- **Design**
- **Distillation**
- **Economics**
- **Energy Balance**
- **Environmental Considerations**
- **Feedstocks**
- **Fermentation**
- **Hardware-Equipment**
- **International**
- **Regulatory**
- **Transportation Use**

GENERAL
Introductory

Brown, M. H., *Brown's Alcohol Motor Fuel Cookbook,* Desert Publications, Cornville, AZ. 86325, 1979, 14 pp. ($9.95).

Commoner, B., *The Politics of Energy,* Alfred A. Knopf, NY, 1979, 102 p.

Crombie, L., *Making Alcohol Fuel—Recipe and Procedure,* 1979, 40 p. Available from Micro-Tech Laboratories, Inc., Route 2, Box #19, Logan, IA 51546. ($4.50).

Gibat and Gibat, *The Lore of Still Building,* 1978, 128 p. Available from Micro Tech Laboratories, Inc., Route 2, Box #19, Logan, IA 51546. ($4.00)

Jawetz, Pincas, "Calculations Linking Farm Policy to Energy Policy," Testimony at Hearings on The Gasohol Motor Fuel Act of 1978, U.S. Senate, Subcommittee on Energy Research and Development, Committee on Energy and Natural Resources, Washington, D.C., August 7-8, 1978, U.S. Government Printing Office, Washington, D.C. 20402, 1978 Publication No. 95-165, pp. 113-142.

National Alcohol Fuels Producers Association, Lincoln, NE, *A Learning Guide for Alcohol Fuel Production,* 1979, 600 p. Available from NAFPA, 2444 B Street, Lincoln, NE 68502, $75.00 with membership.

Nellis, M., *Making it on the Farm. Alcohol Fuel is the Road to Independence,* 1979, 88 p. Available from American Agriculture News, P.O. Box 100, Iredell, TX 76649. ($2.95).

Pimental, D.; et al., "Energy and Land Constraint in Food Production," *Science,* Vol. 190 (No. 4126), November 21, 1975, pp. 754-61.

Reports

Alcohol Fuels Workshop, A Symposium presented for the Farmers Home Administration, Washington, D.C., September 20, 1979.

Third International Symposium on Alcohol Fuels Technology, Asilomar, CA, May 28-31, 1979.

Baratz, B.; Ouellette, R.; Park, W.; Stokes, B.; Mitre Corporation, "Survey of Alcohol Fuel Technology, Volume I," Technical Report No. M74-61-Vol. 1; November, 1975; 43 p. Available from NTIS, PB-256007/6ST, $7.25 paper copy, $3.00 microfiche.

Department of Energy, "Report of the Alcohol Fuels Policy Review," Report No. DOE/PE-0012, June 1979, 119 p. Available from Superintendent of Documents, Stock No. 061-000-00313-4, Washington, D.C.

Freeman, J. H.; et al., American Petroleum Institute, *Alcohols—A Technical Assessment of Their Application as Fuels,* API Publication No. 4261, July 1976.

Office of Technology Assessment, *Gasohol—A Technical Memorandum,* September 1979, 69 p. Available from Superintendent of Documents, U.S. Government Printing Office, Stock No. 052-003-00706-1.

Magazines

Alternative Sources of Energy, Alternative Sources of Energy, Inc., Route 2, Milaca, MN 56353, $6.00/yr. (bimonthly).

BioTimes, The International Biomass Institute, 1522 K Street NW, Suite 600, Washington, D.C. 20005, $10.00 including membership. (monthly).

Farm Show Magazine, 8500 - 210 Street, Lakeville, MN 55044. ($9.00/yr).

Gasohol USA, National Gasohol Commission, P.O. Box 9547, Kansas City, MO 64133, $12.00/yr. (monthly).

Mother Earth News, Mother Earth News, P.O. Box 70, Hendersonville, NC 28739, $15.00/yr. (bimonthly).

Small Farm Energy Project Newsletter, Center for Rural Affairs, P.O. Box 736, Hartington, NE 68739, free. (bimonthly).

Solar Life, SEINAM (Solar Energy Institute), 1110 Sixth Street NW, Washington, D.C. 20001, $5.00/yr. (monthly).

Congressional Hearings

National Fuel Alcohol and Farm Commodity Production Act of 1979, Hearings before the Subcommittees on Conservation and Credit Department Investigations, Oversight and Research, and Livestock and Grains, Committee on Agriculture, U.S. House of Representatives, 96th Congress, May 15-16, 1979. Available from U.S. Government Printing Office, Stock No. 052-070-05071-3.

Oversight—Alcohol Fuel Options and Federal Policies, Hearings before the Subcommittee on Energy Development and Applications of the Committee on Science and Technology, U.S. House of Representatives, 96th Congress, May 4, 1979, June 12, 1979, No. 26—the Committee on Science and Technology.

The Gasohol Motor Fuel Act of 1978, Hearings before the Subcommittee on Energy Research and Development of the Committee on Energy and Natural Resources, U.S. Senate, 95th Congress, August 7-8, 1978, Publication No. 95-165.

Alcohol Fuels, Hearings before the Subcommittee on Advanced Energy Technologies and Energy Conservation Research, U.S. House of Representatives - 95th Congress, July 11-12, 1978. Available from U.S. Government Printing Office (No. 350-520).

Alcohol Fuels, Special Hearing before the Committee on Appropriations, U.S. Senate, 95th Congress, January 31, 1978. Available from U.S. Government Printing Stock Office, No. 052-070-04679-1.

CONVERSION

Bruschke, H., "Direct Processing of Sugarcane into Ethanol," Paper presented at the International Symposium on Alcohol Fuel Technology: Methanol and Ethanol, Wolfsburg, Federal Republic of Germany, November 21-23, 1977. Availability: NTIS, CONF-771175, Complete proceedings, $15.25 printed copy, $3.00 microfiche.

Morrison, Robert T.; Boyd, Robert N., *Organic Chemistry,* 3rd Edition, 1973, Allyn and Bacon, Inc., Rockleigh, NJ.

COPRODUCTS

Distillers Feeds, Distillers Feed Research Council, Cincinnati, OH 45202.

Paturau, J. M., *By-Products of the Cane Sugar Industry,* Elsevier Publishing Company, Amsterdam, 1969, 274 p.

Reilly, P. J., Iowa State University, Ames, IA. "Conversion of Agricultural By-Products to Sugars," Progress report, 1978.

Winston, S. J., Energy Incorporated, "Stillage Treatment Technologies," October 1979, 20 p.

Winston, S. J., Energy Incorporated, "Current State-of-the-Art Stillage Use and Disposal," October 1979, 30 p.

Wisner, R. N.; Gidel, J. O., Iowa Agriculture Experiment Station, "Economic Aspects of Using Grain Alcohol as a Motor Fuel, with Emphasis on By-Product Feed Markets," Report No. 9; June 1977.

DESIGN

Altsheller, W. B.; et al., "Design of a Two-Bushel Per Day Continuous Alcohol Unit," Chemical Engineering Progress, Vol. 43 (No. 9), September 1947, pp. 467-472.

Brackett, A. T.; et al., "Indiana Grain Fermentation Alcohol Plant," 1976, 80 p. Available from Indiana Department of Commerce, State House, Room 336, Indianapolis, IN 46204. (free).

Chambers, R. S., ACR Process Corporation, Urbana, IL, "The Small Fuel-Alcohol Distillery: General Description and Economic Feasibility Workbook," 1979, 21 pages. Available from ACR Process Corporation, 808 S. Lincoln Ave., Urbana, IL 61801. (free).

Katzen, Raphael Associates, "Grain Motor Fuel Alcohol Technical and Economic Assessment Study," Report No. HCP/J6639-01, June 1979, 341 p. Available from NTIS, $12.00 paper copy, $3.00 microfiche.

DISTILLATION

King, C. J., *Separation Processes,* McGraw-Hill, N.Y., 1971.

Tassios, D. P., ed., *Extractive and Azeotropic Distillation,* Advances in Chemistry, Number 115, American Chemical Society, Washington, D.C., 1972.

McCabe, W.; Smith, J. C., *Unit Operations in Chemical Engineering,* Third Edition, McGraw-Hill, NY, 1976.

ECONOMICS

David, M. L.; Hammaker, G. S.; Development Planning and Research Associates, "Gasohol Economic Feasibility Study," Final report, July 1978.

Economics, Statistics, and Cooperative Service, "Gasohol from Grain—The Economic Issues," Final Report AGERSF-21, January 19, 1978, 23 p. Available from NTIS, PB-280120/7ST, $4.00 printed copy, $3.00 microfiche.

Jenkins, D. M., Battelle Columbus Laboratories, "Technical Economic Analysis of the Manufacture of Ethanol from Corn Stover," November 1977.

Jawetz, P., "A New Way to Calculate the Savings from Fuels Substituted for Petroleum," Energy Research Reports, Vol. 5 (No. 18), October 15, 1979.

Koppel, J., Northeast Midwest Institute, Washington, D.C., *Guide to Federal Resources for Economic Development,* September 1979, 137 p.

ENERGY BALANCE

Alich, J. A.; Scholey, F. A.; et al., SRI International. "An Evaluation of the Use of Agricultural Residues as an Energy Feedstock: A Ten Site Survey," Report No. TID-27904/2, January 1978, 402 p. Available from NTIS, $13.25 paper copy, $3.00 microfiche.

Commoner, B., Center for the Biology of Natural Systems, Washington University, St. Louis, MO, "The Potential for Energy Production by U.S. Agriculture," Testimony before the U.S. Senate Committee on Agriculture, Nutrition, and Forestry, Subcommittee on Agricultural Research and General Legislation, July 23, 1979.

Jawetz, P., "Alcohol Additives to Gasoline—An Economic Way for Extending Supplies of Fuel and for Increasing Octane Ratings," American Chemical Society National Meeting, September 9–14, 1979; Division of Petroleum Chemists, Vol. 24 (No. 3), pp. 798–800, Washington, D.C., reprint.

Ladisch, M.R.; Dyck, K., "Dehydration of Ethanol: New Approach Gives Positive Energy Balance," Science, Vol. 205 (No. 31), August 3, 1979, pp. 898–900.

Lewis, C. W., "Fuels from Biomass-Energy Outlays vs. Energy Returns: A Critical Appraisal," Energy, Vol. 2 (No. 3). September 1977, pp. 241–8.

ENVIRONMENTAL CONSIDERATIONS

Brown, D.; McKay, R.; Weir, W., "Some Problems Associated with the Treatment of Effluents from Malt Whiskey Distilleries," Progress in Water Technology, Vol. 8 (No. 2/3), 1976, pp. 291–300.

Jackson, E. A., "Distillery Effluent Treatment in the Brazilian Nationale Alcohol Programme," The Chemical Engineer, April 1977, p. 239–242.

Kant, F. H.; et al., U.S. Environmental Protection Agency, "Feasibility Studies of Alternative Fuels for Automotive Transportation," Report No. EPA–460/374–009, 1974. Available from NTIS, PB235 581/GGI, $4.50 paper copy, $3.00 microfiche.

Lowry, S. O.; Devoto, R. S., Georgia Institute of Technology, "Exhaust Emissions from a Single-Cylinder Engine Fueled with Gasoline, Methanol, and Ethanol," Combustion Science and Technology, Vol. 12 (No. 4, 5, and 6), 1976, pp. 177–182.

FEEDSTOCKS

Chubey, B. B.; Dorrell, D. G., "Jerusalem Artichoke, a Potential Fructose Crop for the Prairies," Journal of the Canadian Institute of Food Science Technology, Vol. 7 (No. 2), 1974, pp. 98–100.

Lipinsky, E. S.; et al., Battelle Columbus Laboratories, "System Study of Fuels from Sugarcane, Sweet Sorghum and Sugar Beets," Vol. 1, Comprehensive Evaluation, Report No. BMI-1957, Vol. 1, March 15, 1977; Vol. 2, Agricultural Considerations, Report No. BMI-1957; Vol. II, December 31, 1976; Vol. 4, Corn Agriculture, Report No. BMI-1957 A; Vol. IV, March 1977.

Nathan, R. A., Battelle Columbus Laboratories, "Fuels from Sugar Crops," Report No. TID–22781, July 1978. Available from NTIS, $8.00 paper copy, $3.00 microfiche.

National Academy of Sciences. Atlas of Nutritional Data on United States and Canadian Feeds. 1971. Available from NAS, 2101 Constitution Ave., Washington, D.C. 20418.

Portola Institute. Energy Primer. 1978.

Potato Alcohol. A Solution to the Energy Crises and Higher Prices for Spuds, Potato Grower of Idaho, April 1978, 50 p.

Stauffer, M.D. Jerusalem Artichoke - Production. Available from Agriculture Canada, Research Station, P.O. Box 3001, Morden Manitoba, ROG IJO, Canada. 1979.

FERMENTATION

Engelbart, W., "Basic Data on Continuous Alcoholic Fermentation of Sugar Solutions and of Mashes from Starch Containing Raw Materials," Paper presented at the International Symposium on Alcohol Fuel Technology: Methanol and Ethanol, Wolfsburg, Federal Republic of Germany, November 21–23, 1977. Available from NTIS, CONF-771175, complete proceedings $15.25 printed copy, $3.00 microfiche.

Lipinsky, E. S.; Scantland, D. A.; McClure, T. A., Battelle Columbus Laboratories, "Systems Study of the Potential Integration of U.S. Corn Prodution and Cattle Feeding with Manufacture of Fuels via Fermentation," Report No. BMI-2033, Vol. I, June 4, 1979. Available from NTIS.

Miller, D. L., Department of Agriculture, Peoria, IL. "Ethanol Fermentation and Potential," Paper presented at the Cellulose Conference in Biotechnological Bioengineering, No. 5, pp . 345, Berkeley, CA, June 25, 1974, USA.

HARDWARE · EQUIPMENT

Lukchis, G. M., "Adsorption Systems: Part 1. Design by Mass-Transfer-Zone Concept," Chemical Engineering, June 11, 1973, pp. 111–116.

Lukchis, G. M., "Adsorption Systems: Part III. Adsorbent Regeneration," Chemical Engineering, August 6, 1973, pp. 83–90.

INTERNATIONAL

Jawetz, P., "The Common Sense Approach in Developing Fuel Alcohols." Workshop on Fermentation Alcohol for Use as Fuel and Chemical Feedstock in Developing Countries, Vienna, Austria, March 26–30,

1979, Paper no. ID/WG.293/12, United Nations Industrial Development Organization. (Papers dated: February 5, 1979—abstract, March 26, 1979—paper).

Ribeiro, Filho F. A., "The Ethanol-Based Chemical Industry in Brazil," Workshop on Fermentation Alcohol for Use as Fuel and Chemical Feedstock in Developing Countries, Vienna, Austria, March 1979, Paper No. ID/WG.293/4 UNIDO, United Nations Industrial Development Organization.

Sharma, K. D., "Present Status of Alcohol and Alcohol Based Chemical Industry in India," Workshop on Fermentation Alcohol for Use as Fuel and Chemical Feedstock in Developing Countries, Vienna, Austria, March 26–30, 1979, paper no. ID/WG.293/14 UNIDO, United Nations International Development Organization.

REGULATORY

Abeles, T. P.; King, Janna R., Minnesota Legislature Science and Technology Project, "Parameters for Legislature Consideration of Bioconversion Technologies," February 1978, 45 p. Available from NTIS, PB 284742/45T, $4.50 paper copy, $3.00 microfiche.

Bureau of Alcohol, Tobacco, and Firearms, *Ethyl Alcohol for Fuel Use*, Informational Brochure, ATF PX 5000.1, July 1978, 7 p.

Bureau of Alcohol, Tobacco, and Firearms, *Alcohol Fuel and ATF*, Informational Brochure, ATF P5000.2, August 1979, 4 p.

TRANSPORTATION USE

Adt, R. R., Jr., et al., University of Miami, "Effects of Blending Ethanol with Gasoline on Automotive Engines' Steady State Performance and Regulated Emissions Characteristics," Topical Report, January 1978.

Allsup, J. R.; Eccleston, D. B., "Ethanol/Gasoline Blends as Automotive Fuel," Report No. BETC/RI-79/2, May 1979, 13 p. Available from NTIS, $4.00 paper copy, $3.00 microfiche.

Bernhardt, W., "Future Fuels and Mixture Preparation Methods for Spark Ignition Automobile Engines," Progress in Energy and Combustion Science, Vol. 3 (No. 3), 1977, pp. 139–150.

Bushnell, D. J.; Simonsen, J. M., "Alcohol Assisted Hydrocarbon Fuels: A Comparison of Exhaust Emissions and Fuel Consumption Using Steady-State and Dynamic Engine Test Facilities," Energy Communications, Vol. 2 (No. 2), 1976, pp. 107–132.

Panchapakesan, N. R.; Gopalakrishnan, K. V., "Factors that Improve the Performance of an Ethanol-Diesel Oil Dual-Fuel Engine," Paper presented at the International Symposium on Alcohol Fuel Technology: Methanol and Ethanol, Wolfsburg, Federal Republic of Germany, November 21–23, 1977. Available from NTIS, CONF-771175, complete proceedings $15.25 printed copy, $3.00 microfiche.

Scott, W. M., Ricardo Consulting Engineers, "Alternative Fuels for Automotive Diesel Engines," Paper presented at the Symposium on Future Automotive Fuels-Prospects, Performance, and Perspective, Warren, MI, October 6, 1975, in *Future Automotive Fuels: Prospects, Performance, and Perspective*, Colucci, J. M.; Gallopoolos, N. E. (eds.); pp. 263–292. Available from NTIS, CONF-751018.

APPENDIX D

Glossary

ACID HYDROLYSIS: decomposition or alteration of a chemical substance by acid.

ACIDITY: the measure of how many hydrogen ions a solution contains.

AFLATOXIN: the substance produced by some strains of the fungus *Aspersillus Flavus;* the most potent carcinogen yet discovered; a persistent contaminant of corn that renders crops unsalable.

ALCOHOL: the family name of a group of organic chemical compounds composed of carbon, hydrogen, and oxygen; a series of molecules that vary in chain length and are composed of a hydrocarbon plus a hydroxyl group, $CH_3-(CH_2)n-OH$; includes methanol, ethanol, isopropyl alcohol, and others.

ALDEHYDES: any of a class of highly reactive organic chemical compounds obtained by oxidation of primary alcohols, characterized by the common group CHO, and used in the manufacture of resins, dyes, and organic acids.

ALKALI: soluble mineral salt of a low density, low melting point, highly reactive metal; characteristically "basic" in nature.

ALPHA-AMYLASE - AMYLASE: enzyme which converts starch into sugars.

AMBIENT: the prevalent surrounding conditions usually expressed as functions of temperature, pressure, and humidity.

AMINO ACIDS: the naturally-occurring, nitrogen-containing building blocks of protein.

AMYLODEXTRINS: see Dextrins.

ANAEROBIC DIGESTION: without air; a type of bacterial degradation of organic matter that occurs only in the absence of air (oxygen).

ANHYDROUS: a compound that does not contain water either absorbed on its surface or as water of crystallization.

ATMOSPHERIC PRESSURE: pressure of the air (and atmosphere surrounding us) which changes from day to day; it is equal to 14.7 psia.

AZEOTROPE: the chemical term for two liquids that, at a certain concentration, boil at the same temperature; alcohol and water cannot be separated further than 194.4 proof because at this concentration, alcohol and water form an azeotrope and vaporize together.

AZEOTROPIC DISTILLATION: distillation in which a substance is added to the mixture to be separated in order to form an azeotropic mixture with one or more of the components of the original mixture; the azeotrope formed will have a boiling point different from the boiling point of the original mixture which will allow separation to occur.

BALLING HYDROMETER OR BRIX HYDROMETER: a triple-scale wine hydrometer designed to record the specific gravity of a solution containing sugar.

BARREL: a liquid measure equal to 42 American gallons or about 306 pounds; one barrel equals 5.6 cubic feet or 0.159 cubic meters; for crude oil, one barrel is about 0.136 metric tons, 0.134 long tons, and 0.150 short tons.

BASIC HYDROLYSIS: decomposition or alteration of a chemical substance by alkali (basic) solution.

BATCH FERMENTATION: fermentation conducted from start to finish in a single vessel.

BATF: Bureau of Alcohol, Tobacco, and Firearms; under the U.S. Department of Treasury. Responsible for the issuance of permits, both experimental and commercial, for the production of alcohol.

BEER: the product of fermentation by microorganisms; the fermented mash, which contains about 11-12% alcohol; usually refers to the alcohol solution remaining after yeast fermentation of sugars.

BEER STILL: the stripping section of a distillation column for concentrating ethanol.

BEER WELL: the surge tank used for storing beer prior to distillation.

BETA - AMYLASE: see Amylase.

BIOMASS: plant material, includes cellulose carbohydrates, ligniferous constituents, etc.

BOILING POINT: the temperature at which the transition from the liquid to the gaseous phase occurs in a pure substance at fixed pressure.

BRITISH THERMAL UNIT (Btu): the amount of heat required to raise the temperature of one pound of water one degree Fahrenheit under stated conditions of pressure and temperature (equal to 252 calories, 778 foot-pounds, 1,055 joules, and 0.293 watt-hours); it is the standard unit for measuring quantity of heat energy.

BULK DENSITY: the mass (weight) of a material divided by the actual volume it displaces as a whole substance expressed in lb/ft³; kg/m³; etc.

CALORIE: the amount of heat required to raise one gram of water one degree centigrade.

CARBOHYDRATE: a chemical term describing compounds made up of carbon, hydrogen, and oxygen; includes all starches and sugars.

CARBON DIOXIDE: a gas produced as a by-product of fermentation; chemical formula is CO_2.

CASSAVA: a starchy root crop used for tapioca; can be grown on marginal croplands along the southern coast of the United States.

CELL RECYCLE: the process of separating yeast from fully fermented beer and returning it to ferment a new mash; can be done with clear worts in either batch or continuous operations.

CELLULASE: an enzyme capable of splitting cellulose.

CELLOUSE: the main polysaccharide in living plants, forms the skeletal structure of the plant cell wall; can be hydrolyzed to glucose.

CELSIUS (Centigrade): a temperature scale commonly used in the sciences; at sea level, water freezes at 0 °C and boils at 100 °C.

CENTRIFUGE: a rotating device for separating liquids of different specific gravities or for separating suspended colloidal particles according to particle-size fractions by centrifical force.

CHLOROPLAST: a small portion of a plant cell which contains the light-absorbing pigment chlorophyll, and converts light energy to chemical energy.

COLUMN: vertical, cylindrical vessel used to increase the degree of separation of liquid mixtures by distillation or extraction.

COMPOUND: a chemical term denoting a combination of two or more distinct elements.

CONCENTRATION: ratio of mass or volume of solute present in a solution to the amount of solvent.

CONDENSER: a heat-transfer device that reduces a thermodynamic fluid from its vapor phase to its liquid phase.

CONTINUOUS FERMENTATION: a steady-state fermentation system that operates without interruption; each stage of fermentation occurs in a separate section of the fermenter, and flow rates are set to correspond with required residence times.

COOKER: a tank or vessel designed to cook a liquid or extract or digest solids in suspension; the cooker usually contains a source of heat; and is fitted with an agitator.

COPRODUCTS: the resulting substances and materials that accompany the production of ethanol by fermentation process.

DDGS: see Distiller Dried Grains with Solubles.

DEHYDRATION: the removal of 95% or more of the water from any substance by exposure to high temperature.

DENATURANT: a substance that makes ethanol unfit for human consumption.

DENATURE: the process of adding a substance to ethyl alcohol to make it unfit for human consumption; the denaturing agent may be gasoline or other substances specified by the Bureau of Alcohol, Tobacco, and Firearms.

DEPARTMENT OF ENERGY: in October 1977, the Department of Energy (DOE) was created to consolidate the multitude of energy-oriented government programs and agencies; the Department carries out its mission through a unified organization that coordinates and manages energy conservation, supply development, information collection and analysis, regulation, research, development, and demonstration.

DESICCANT: a substance having an affinity for water; used for drying purposes.

DEWATERING: to remove the free water from a solid substance.

DEXTRINS: a polymer of D-Glucose which is intermediate in complexity between starch and maltose formed by hydrolysis of starches.

DEXTROSE: the same as glucose.

DISACCHARIDES: the class of compound sugars which yield two monosaccharide units upon hydrolysis; examples are sucrose, mannose, and lactose.

DISPERSION: the distribution of finely divided particles in a medium.

DISTILLATE: that portion of a liquid which is removed as a vapor and condensed during a distillation process.

DISTILLATION: the process of separating the components of a mixture by differences in boiling point; a vapor is formed from the liquid by heating the liquid in a vessel and successively collecting and condensing the vapors into liquids.

DISTILLERS DRIED GRAINS (DDG): the dried distillers grains by-product of the grain fermentation process which may be used as a high-protein (28%) animal feed. (See distillers grains.)

DISTILLERS DRIED GRAINS WITH SOLUBLES (DDGS): a grain mixture obtained by mixing distillers dried grains and distillers dried solubles.

DISTILLERS DRIED SOLUBLES (DDS): a mixture of water-soluble oils and hydrocarbons obtained by condensing the thin stillage fraction of the solids obtained from fermentation and distillation processes.

DISTILLERS FEEDS: primary fermentation products resulting from the fermentation of cereal grains by the yeast *Saccharomyces cerevisiae*.

DISTILLERS GRAIN: the nonfermentable portion of a grain mash comprised of protein, unconverted carbohydrates and sugars, and inert material.

ENRICHMENT: the increase of the more volatile component in the condensate of each successive stage above the feed plate.

ENSILAGE: immature green forage crops and grains which are preserved by alcohol formed by an anaerobic fermentation process.

ENZYMES: the group of catalytic proteins that are produced by living microorganisms; enzymes mediate and promote the chemical processes of life without themselves being altered or destroyed.

ETHANOL: C_2H_5OH; the alcohol product of fermentation that is used in alcohol beverages and for industrial purposes; chemical formula blended with gasoline to make gasohol; also known as ethyl alcohol or grain alcohol.

ETHYL ALCOHOL: also known as ethanol or grain alcohol; see Ethanol.

EVAPORATION: conversion of a liquid to the vapor state by the addition of latent heat of vaporization.

FACULTATIVE (ANAEROBE): a microorganism that grows equally well under aerobic and anaerobic conditions.

FAHRENHEIT SCALE: a temperature scale in which the boiling point of water is 212° and its freezing point 32°; to convert °F to °C, subtract 32, multiply by 5, and divide the product by 9 (at sea level).

FEED PLATE: the theoretical position in a distillation column above which enrichment occurs and below which stripping occurs.

FEEDSTOCK: the base raw material that is the source of sugar for fermentation.

FERMENTABLE SUGAR: sugar (usually glucose) derived from starch and cellulose that can be converted to ethanol (also known as reducing sugar or monosaccharide).

FERMENTATION: a microorganically mediated enzymatic transformation of organic substances, especially carbohydrates, generally accompanied by the evolution of a gas.

FERMENTATION ETHANOL: ethyl alcohol produced from the enzymatic transformation of organic substances.

FLASH HEATING: very rapid heating of material by exposure of small fractions to temperature and using high flow rates.

FLASH POINT: the temperature at which a combustible liquid will ignite when a flame is introduced; anhydrous ethanol will flash at 51° F, 90-proof ethanol will flash at 78° F.

FLOCCULATION: the aggregation of fine suspended particles to form floating clusters or clumps.

FOSSIL FUEL: any naturally occurring fuel of an organic nature such as coal, crude oil, or natural gas.

FRACTIONAL DISTILLATION: a process of separating alcohol and water (or other mixtures).

FRUCTOSE: a fermentable monosaccharide (simple) sugar of chemical formula $C_6H_{12}O_6$. Fructose and glucose are optical isomers; that is, their chemical structures are the same but their geometric configurations are mirror images of one another.

FUSEL OIL: a clear, colorless, poisonous liquid mixture of alcohols obtained as a byproduct of grain fermentation; generally amyl, isoamyl, propyl, isopropyl, butyl, isobutyl alcohols and acetic and lactic acids.

GASOHOL (Gasahol): registered trade names for a blend of 90% unleaded gasoline with 10% fermentation ethanol.

GASOLINE: a volatile, flammable liquid obtained from petroleum that has a boiling range of approximately 29°–216° C and is used as fuel for spark-ignition internal combustion engines.

GELATINIZATION: the rupture of starch granules by temperature which forms a gel of soluble starch and dextrins.

GLUCOSE: a monosaccharide; occurs free or combined and is the most common sugar; chemical formula $C_6H_{12}O_6$.

GLUCOSIDASE: an enzyme that hydrolyzes any polymer of glucose monomers (glucoside). Specific glucosidases must be used to hydrolyze specific glucosides; e.g., B-glucosidases are used to hydrolyze cellulose; α-glucosidases are used to hydrolyze starch.

GRAIN ALCOHOL: see Ethanol.

HEAT EXCHANGER: a device that transfers heat from one fluid (liquid or gas) to another, or to the environment.

HEAT OF CONDENSATION: the same as the heat of vaporization, except that the heat is given up as the vapor condenses to a liquid at its boiling point.

HEAT OF VAPORIZATION: the heat input required to change a liquid at its boiling point (water at 212° F) to a vapor at the same temperature (212° F).

HEATING VALUE: the amount of heat obtainable from a fuel and expressed, for example, in Btu/lb.

HEXOSE: any of various simple sugars that have six carbon atoms per molecule.

HYDRATED: chemically combined with water.

HYDROCARBON: a chemical compound containing hydrogen, oxygen, and carbon.

HYDROLYSIS: the decomposition or alteration of a polymeric substance by chemically adding a water molecule to the monomeric unit at the point of bonding.

HYDROMETER: a long-stemmed glass tube with a weighted bottom; it floats at different levels depending on the relative weight (specific gravity) of the liquid; the specific gravity of other information is read where the calibrated stem emerges from the liquid.

INDOLENE: a chemical used in comparative tests of automotive fuels.

INOCULUM: a small amount of bacteria produced from a pure culture which is used to start a new culture.

INULIN: a polymeric carbohydrate comprised of fructose monomers found in the roots of many plants, particularly Jerusalem artichokes.

LACTIC ACID: $C_3H_6O_3$, the acid formed from milk sugar (lactose) and produced as a result of fermentation of carbohydrates by bacteria called *Lactobaccilus*.

LACTOSE: a white crystalline disaccharide made from whey and used in pharmaceuticals, infant foods, bakery products, and confections; also called "milk sugar", $C_{12}H_{22}O_{11}$.

LEADED GASOLINE: gasoline containing tetraethyllead to raise octane value.

LIGNIFIED CELLULOSE: cellulose polymer wrapped in a polymeric sheath extremely resistant to hydrolysis because of the strength of its linkages called lignin.

LINKAGE: the bond or chemical connection between constituents of a polymeric molecule.

LIQUEFACTION: the change in the phase of a substance to the liquid state; in the case of fermentation, the conversion of water-insoluble carbohydrate to water-soluble carbohydrate.

MALT: barley softened by steeping in water, allowed to germinate, and used especially in brewing and distilling.

MASH: a mixture of grain and other ingredients with water to prepare wort for brewing operations.

MEAL: a granular substance produced by grinding.

MEMBRANE: a sheet polymer capable of separating liquid solutions.

METHANOL: a light volatile, flammable, poisonous, liquid alcohol, CH_3OH, formed in the destructive distillation of wood or made synthetically and used especially as a fuel, a solvent, an antifreeze, or a denaturant for ethyl alcohol, and in the synthesis of other chemicals; methanol can be used as fuel for motor vehicles; also known as methyl alcohol or wood alcohol.

METHYL ALCOHOL: also known as methanol or wood alcohol; see Methanol.

MOLECULAR SEIVE: a column which separates molecules by selective adsorption of molecules on the basis of size.

MOLECULE: the chemical term for the smallest particle of matter that is the same chemically as the whole mass.

MONOMER: a simple molecule which is capable of combining with a number of like or unlike molecules to form a polymer.

MONOSACCHARIDES: see Fermentable Sugar.

OCTANE NUMBER: a rating which indicates the tendency to knock when a fuel is used in a standard internal combustion engine under standard conditions.

OSMOTIC PRESSURE: osmosis - applied pressure required to prevent passage of a solvent across a membrane which separates solutions of different concentrations.

OSMOPHYLLIC: organisms which prosper in solutions with high osmotic pressure.

PACKED DISTILLATION COLUMN: a column or tube constructed such that a packing of ceramics, steel, copper, or fiberglass-type material.

pH: a term used to describe the free hydrogen ion concentration of a system; a solution of pH 0 to 7 is acid; pH of 7 is neutral; pH over 7 to 14 is alkaline.

PLATE DISTILLATION COLUMN (Sieve tray column): a distillation column constructed with perforated plates or screens.

POLYMER: a substance made of molecules comprised of long chains or cross-linked simple molecules.

POUNDS PER SQUARE INCH ABSOLUTE (psia): the measurement of pressure referred to a complete vacuum or 0 pressure.

POUNDS PER SQUARE INCH GUAGE (psig): expressed as a quantity measured from above atmospheric pressure.

POUND OF STEAM: one pound (mass) of water in the vapor phase not to be confused with the steam *pressure* which is expressed in *pounds per square inch.*

PRACTICAL YIELD: the amount of product that can actually be derived under normal operating conditions; i.e., the amount of sugar that normally can be obtained from a given amount of starch or the amount of alcohol that normally can be obtained is usually less than theoretical yield.

PROOF: a measure of ethanol content; 1 percent equals 2 proof.

PROOF GALLON: a U.S. gallon of liquid which is 50% ethyl alcohol by volume; also one tax gallon.

PROTEIN: any of a class of high molecular weight polymer compounds comprised of a variety of amino acids joined by a peptide linkage.

PYROLYSIS: the breaking apart of complex molecules into simpler units by heating in the absence of stoichiometric quantities of oxygen.

QUAD: one quadrillion (10^{15} or 1,000,000,000,000,000) Btu's (British thermal units).

RECTIFICATION: with regard to distillation, the selective increase of the concentration of the lower volatile component in a mixture by successive evaporation and condensation.

RECTIFYING COLUMN: the portion of a distillation column above the feed tray in which rising vapor is enriched by interaction with a countercurrent falling stream of condensed vapor.

REFLUX: that part of the product stream that may be returned to the process to assist in giving increased conversion or recovery.

RELATIVE DENSITY: see Specific Gravity.

RENEWABLE RESOURCES: renewable energy; resources that can be replaced after use through natural means; example: solar energy, wind energy, energy from growing plants.

ROAD OCTANE: a numerical value for automotive anti-knock properties of a gasoline; determined by operating a car over a stretch of level road.

SACCHARIFY: to hydrolyze a complex carbohydrate into a simpler soluble fermentable sugar, such as glucose or maltose.

SACCHAROMYCES: a class of single-cell yeasts which selectively consume simple sugars.

SCRUBBING EQUIPMENT: equipment for countercurrent liquid-vapor contact of flue gases to remove chemical contaminants and particulates.

SETTLING TIME: in a controlled system, the time required for entrained or colloidal material to separate from the liquid.

SIGHT GAUGE: a clear calibrated cylinder through which liquid level can be observed and measured.

SIMPLE SUGARS: see Fermentable Sugars.

SOLAR ENERGY RESEARCH INSTITUTE (SERI): the Solar Energy Research Development and Demonstration Act of 1974 called for the establishment of SERI, whose general mission is to support DOE's solar energy program and foster the widespread use of all aspects of solar technology, including direct solar conversion (photovoltaics), solar heating and cooling, solar thermal power generation, wind conversion, ocean thermal conversion, and biomass conversion.

SPECIFIC GRAVITY: the ratio of the mass of a solid or liquid to the mass of an equal volume of distilled water at 4°C.

SPENT GRAINS: the nonfermentable solids remaining after fermentation of a grain mash.

STARCH: a carbohydrate polymer comprised of glucose monomers linked together by a glycosidic bond and organized in repeating units; starch is found in most plants and is a principal energy storage product of photosynthesis; starch hydrolyzes to several forms of dextrin and glucose.

STILL: an apparatus for distilling liquids, particularly alcohols; it consists of a vessel in which the liquid is vaporized by heat, and a cooling device in which the vapor is condensed.

STILLAGE: the nonfermentable residue from the fermentation of a mash to produce alcohol.

STOVER: the dried stalks and leaves of a crop remaining after the grain has been harvested.

STRIPPING SECTION: the section of a distillation column below the feed in which the condensate is progressively decreased in the fraction of more volatile component by stripping.

SUCROSE: a crystalline disaccharide carbohydrate found in many plants, mainly sugar cane, sugar beets, and maple trees; $C_{12}H_{22}O_{11}$.

THERMOPHYLLIC: capable of growing and surviving at high temperatures.

THIN STILLAGE: the water-soluble fraction of a fermented mash plus the mashing water.

VACUUM DISTILLATION: the separation of two or more liquids under reduced vapor pressure; reduces the boiling points of the liquids being separated.

VAPORIZE: to change from a liquid or a solid to a vapor, as in heating water to steam.

VAPOR PRESSURE: the pressure at any given temperature of a vapor in equilibrium with its liquid or solid form.

WHOLE STILLAGE: the undried "bottoms" from the beer well comprised of nonfermentable solids, distillers solubles, and the mashing water.

WOOD ALCOHOL: see Methanol.

WORT: the liquid remaining from a brewing mash preparation following the filtration of fermentable beer.

YEAST: single-cell microorganisms (fungi) that produce alcohol and CO under anaerobic conditions and acetic acid and CO under aerobic conditions; the microorganism that is capable of changing sugar to alcohol by fermentation.

ZYMOSIS: see Fermentation.

Department of Treasury Bureau of Alcohol, Tobacco, and Firearms

Permit Information

- Alcohol Fuels and BATF
- Regulatory Requirements
- BATF Regional Offices
- Sample Application for Experimental Distilled Spirits Plant

ALCOHOL FUELS AND BATF

The Bureau of Alcohol, Tobacco, and Firearms (BATF) is responsible for administering federal laws and regulations governing the taxation, production, denaturation, and distribution of fermentation ethanol. They have the responsibility to make available, in an understandable format, the information required by prospective ethanol producers and users so that they can comply with current laws and regulations.

The information required to qualify a distilled spirits plant for commercial production of ethanol has been summarized in BATF publication 5000.1, which is available at the BATF Regional Office. This convenient publication provides the applicable regulations and pertinent information to prospective producers and distributors of ethanol for commercial fuel ventures. Many citizens, however, are not interested in going into the fuel business; but instead, want to produce fermentation ethanol as a supplemental fuel for their own use. They want to make themselves more self-sufficient and reduce their fuel bills. This section will provide the information needed to qualify with BATF requirements to produce ethanol for individual use.

Who Can Qualify?

Under Title 27, Code of Federal Regulations, Section 201.64, **anyone** may establish an experimental distilled spirits plant for experimentation in, or development of:

- sources of material from which spirits may be produced;

- processes by which spirits may be produced or refined; or

- industrial uses of spirits.

Since "home production" of ethanol for use as an alternate fuel is a new concept, the BATF can and will approve the establishment of experimental plants by individuals who are seeking to prove the merit of this idea. There is no BATF application form to fill out; a letter written by the individual with certain pertinent information is presented. The information required in this letter includes natur and purpose, description of plant premises, description of production process and equipment, security, and rate production. Explanations of these items follow.

Nature and Purpose

A general statement describing what is intended is required. For example, the applicant wants to produce fermentation ethanol which will be used as fuel for farm engines and heaters; including, but not limited to, tractors, combines, swathers, cars, trucks, irrigation motors, houses, grain dryers, etc.; or, the applicant intends to experiment with waste products (corn cobs, stalks, spoiled grain) to produce fermentation ethanol to determine if it can be refined and used as a fuel to run farm implements; or, the applicant intends to build a still which can efficiently refine fermentation ethanol produced from waste products for ultimate use as a fuel.

Description of Plant Premises

Describe the location of the premises where the experimental plant will be established. If the applicant is a farmer, this should include the entire farm so that the ethanol produced may be used without removing it from the "plant premises." The description should include the number of acres involved. The buildings used in the production and storage of ethanol (if applicable) and their relative location on the farm should be described.

Description of Production Process and Equipment

The production process to be used should be described. For example, "mash to be fermented will consist of spoiled grain, vegetables, and kitchen garbage. Initial distillation of fermentation ethanol will be accomplished by solar energy. Ethanol recovered from the mash via the still will be refined in a still of my own manufacture." A descriptive list of the equipment used in the process should be given. All equipment from the mash tank to the finished ethanol storage tank should be included. In addition, when and how the ethanol produced will be denatured should be described; i.e., "ethanol will be mixed with gasoline in a 10:1 ratio immediately after production. This will be done in the fuel storage tank."

Security

Security measures to be provided for the ethanol produced should be described. For example, "ethanol will be stored in a 1,500-gallon fuel tank which is equipped for locking with a padlock; or, ethanol will be immediately denatured, drummed off, then stored in a locked shed. All windows in the shed are equipped with security screens and two watchdogs are on the premises."

Rate of Production

The amount of ethanol expected to be produced in an average 15-day period must be stated. This may be estimated. The average proof of the finished ethanol produced must be given. For example, "approximately 400 gallons of ethanol, averaging between 160–190 proof, will be produced in a 15-day period."

The completed letter application must be filed with the Regional BATF Office. They will examine it for completeness then forward it to one of their field officers for inspection. A BATF inspector will visit the proposed plant prior to the time *formal* authorization to operate is given. The facilities will be examined and the applicant will be advised of the record requirements for the operation. The inspector will also discuss the authorized operations and answer any questions. The inspector will supply blank bond forms (F-2601) and inform the applicant of the dollar amount of coverage required. [Form 2601 is a surety bond which is executed by the applicant as the principal and an approved insurance company as the surety. The amount of the bond will depend upon the type and volume of operations to be conducted. It is the BATF estimate of the *potential* tax liability on the spirits produced. The rate of tax liability is $10.50 per proof gallon (1 gallon of 100-proof spirits). No tax is actually paid for ethanol produced or used on the experimental distilled spirits plant. However, the bond coverage is still necessary to ensure the protection of the tax liabilities which attach to all spirits produced.] See the sample Form 2607 which follows.

The BATF inspector will also deliver Forms 4805 and 4871 (relating to water quality considerations and environmental information, respectively) which can be completed and returned to the BATF inspector during his/her visit. After the BATF Regional Office receives a favorable inspection report and a properly executed bond, they will issue a formal authorization to begin operations. This authorization, however, does not exempt the applicant from complying with any state and local requirements concerning the production of fermentation ethanol.

The following questions and answers are presented to clarify certain points relative to the experimental distilled spirits plant application and operation.

1. **Question:** Can I sell or loan any excess ethanol produced to another person for fuel use?

   Answer: No. The ethanol produced may be used as fuel only at the plant premises described in your bond and letter application. Only a plant qualified as a commercial distilled spirits plant (DSP) can sell, loan, or give ethanol to another party.

2. **Question:** Can I remove some of the ethanol from the plant premises for my own use? (For example, as a fuel for my personal car.)

   Answer: You may remove ethanol from your plant premises for your own use as a fuel; however, the ethanol must be *completely* denatured according to one of the two formulas listed below before removal.

   Formula No. 18. To every 100 gallons of ethanol add:
   2.5 gallons of methyl isobutyl ketone;

0.125 gallon of pyronate or a compound similar thereto;
0.50 gallon of acetaldol; and
1 gallon of either kerosene or gasoline.

or

Formula No. 19. To every 100 gallons of ethanol add:
4.0 gallons of methyl isobutyl ketone; and
1.0 gallon of either kerosene or gasoline.

3. **Question:** Must I denature the ethanol before using it on my farm?

   Answer: While BATF can approve applications where good cause is shown for the need to use undenatured spirits as fuel, they prefer that you denature your ethanol with gasoline, diesel fuel, or heating fuel immediately after production. This will allow BATF to approve less stringent security systems and recordkeeping requirements than what they impose on applicants who do not denature their ethanol.

4. **Question:** Can any of the ethanol produced be used for beverage purposes?

   Answer: Absolutely not. Besides the IRS Excise Tax of $10.50 per proof gallon, for which you would become liable, you could also incur severe criminal penalties.

5. **Question:** Can I build my distillery system prior to receiving any authorization from BATF to operate?

   Answer: Yes. However, you must file an application to establish an experimental distilled spirits plant with BATF immediately after its completion. You may, however, file sooner if you feel you will have the equipment set up prior to the qualification visit by the BATF inspector. However, under no circumstances may you start producing spirits prior to receipt of a formal authorization by the BATF.

6. **Question:** Can I qualify two or more farms for my plant premises?

Answer: Yes, if they are in close proximity to each other so as to allow a BATF inspector to inspect all premises without causing undue travel and administrative difficulties.

7. **Question:** How long a period is the authorization effective?

   Answer: BATF is currently approving operations for a 2-year period unless you include in your application some justification for a longer period of time.

8. **Question:** Can I renew my experimental plant authorization after expiration?

   Answer: Yes. When the authorization expires, you may file a new application listing the current information on all subjects originally described (security, rate of production, equipment, etc.). A new bond will not be required unless significant changes in operations have occurred since the original filing.

9. **Question:** Can partnerships and corporations make application as well as individuals?

   Answer: Yes, however, if the application is filed by a partnership, all partners must sign it. If it is filed by a corporation, a person authorized by the corporation must sign and proof of such authorization must accompany the application (e.g., a certified copy of a corporate resolution or an abstract of bylaws giving such authority).

10. **Question:** Can I convert to a commercial operation?

Answer: There is no simple means of converting an experimental operation into a commercial operation. Normally, all provisions of Title 26, U.S.C. Chapter 51 and Title 27, CFR Part 201 will be waived for an experimental ethanol fuel-related DSP except for those relating to:

(1) filing application for, and receiving approval to operate an experimental DSP for a limited, specified period of time;

(2) filing of a surety bond to cover the tax on the ethanol produced;

(3) attachment, assessment, and collection of tax;

(4) authorities of BATF Officers; and

(5) maintenance of records.

No such blanket waiver will be given for commercial operations. The applicant will have to follow the qualification procedure outlined in BATF P 5000.1. BATF will, however, give favorable consideration to alternate procedures from regulations which do not present a definite jeopardy to the revenue; however, each such variation will be viewed and ruled upon on an individual basis.

Hopefully, changes in the law and regulations to facilitate the qualification of commercial plants will occur within the next 2 years. BATF also has proposed simpler permit applications for small-scale fermentation ethanol plants producing fuel; however, they have not yet been approved.

REGULATORY REQUIREMENTS

The Specific Regulatory Requirements of the Bureau of Alcohol, Tobacco, and Firearms are provided on the following pages.

Applicable Excerpts From
27 CFR Part 201—
Distilled Spirits Plants

Subpart F—Qualifications of Distilled Spirits Plants

§201.131 General requirements for registration.

A person shall not engage in the business of a distiller, bonded warehouseman, rectifier, or bottler of distilled spirits, unless he has made application for and has received notice of registration of his plant with respect to such business as provided in this part. Application for registration shall be made on Form 2607 to the assistant regional commissioner. Each application shall be executed under penalties of perjury, and all written statements, affidavits, and other documents submitted in support of the application or incorporated by reference shall be deemed to be a part thereof. The assistant regional commissioner may, in any instance where the outstanding notice of registration is inadequate or incorrect in any respect, require by registered or certified mail the filing of an application on Form 2607 to amend the notice of registration, specifying the respects in which amendment is required. Within 60 days after the receipt of such notice, the proprietor shall file such application.

(72 Stat. 1349; 26 U.S.C. 5171, 5172)

§201.132 Data for application for registration.

Application on Form 2607 shall be prepared in accordance with the headings on the form, and instructions thereon and issued in respect thereto, and shall include the following:

(a) Serial number and statement of purpose for which filed.

(b) Name and principal business address of the applicant, and the location of the plant if different from the business address.

(c) Statement of the type of business organization and of the

persons interested in the business, supported by the items of information listed in §201.148.

(d) Statement of the business or businesses to be conducted.

(e) In respect of the plant to which the Form 2607 relates, a list of applicant's operating and basic permits, and of the qualification bonds (including those filed with the application) with the name of the surety or sureties for each bond.

(f) List of the offices, the incumbents of which are authorized by the articles of incorporation or the board of directors to act on behalf of the proprietor or to sign his name.

(g) Plat and plans (see §§201.154–201.159).

(h) Description of the plant (see §201.149).

(i) List of major equipment (see §201.147).

(j) As applicable, the following:

(1) With respect to the business of a distiller:

(i) Statement of maximum proof gallons that will be (*a*) produced during a period of 15 days and (*b*) in transit to the bonded premises. (Not required if the qualification bond is in the maximum sum.)

(ii) Statement of daily producing capacity in proof gallons.

(iii) Statement of process (see §201.153).

(iv) Statement whether denaturing operations will be conducted.

(v) Statement of title to the bonded premises and interest in the equipment used for the production of spirits, accompanied where required by consent on Form 1602 (see §§201.151–201.152).

(2) With respect to the business of a bonded warehouseman:

(i) Statement of the maximum proof gallons that will be stored on, and in transit to, the bonded premises. (Not required if the qualification

bond is in the maximum sum.)

(ii) Description of the system of storage, and statement of storage capacity (bulk, packages, and cases).

(iii) Statement whether denaturing and/or bottling-in-bond operations will be conducted.

(3) With respect to the business of a rectifier, a statement of the maximum tax the rectifier will be liable to pay under sections 5021 and 5022, I.R.C., in a 30-day period. (Not required if the qualification bond is in the maximum sum.)

(4) With respect to the business of bottling after tax determination:

(i) Statement of the name, address, and plant number of a plant qualified by the applicant for production or bonded warehousing. (Not required if the plant being registered is qualified for production, bonded warehousing, or rectification, or if the applicant is a State or political subdivision thereof.)

(ii) Statement whether operations involving bottling in bond after tax determination, as provided in § 201.114, will be conducted.

(5) With respect to any other business to be conducted on the plant premises, as provided by Subpart D of this part, a description of such business, a list of the buildings and/or equipment to be used, and a statement as to the relationship, if any, of such business to distilled spirits operations at the plant.

Where any of the information required by paragraph (c) or paragraph (g) of this section is on file with the assistant regional commissioner, such information, if accurate and complete, may, by incorporation by reference thereto by the applicant, be made a part of the application for registration. The applicant shall, when so required by the assistant regional commissioner, furnish as a part of his application for registration such additional information as may be necessary for the assistant regional commissioner to determine whether the application for registration should be approved.

(72 Stat. 1349; 26 U.S.C. 5171, 5172)

[25 FR 6053, June 30, 1960, as amended by T.D. 6749, 29 FR 9896, July 23, 1964; T.D. 7112, 36 FR 8571, May 8, 1971]

§201.133 Notice of registration.

The application for registration, when approved, shall constitute the notice of registration of the plant. A plant shall not be registered or reregistered under this subpart until the applicant has complied with all requirements of law and regulations relating to the qualification of the business or businesses in which the applicant intends to engage. A plant shall not be operated unless the proprietor has a valid notice of registration covering the businesses and operations to be conducted at such plant. In any instance where a bond is required to be given or a permit is required to be obtained with respect to a business or operation before notice of registration of the plant may be received with respect thereto, the notice of registration shall not be valid with respect to such business or operation in the event that such bond or permit is no longer in effect and an application for reregistration shall be filed and notice of registration again obtained before thereafter engaging in such business or operation at such plant; reregistration is not required when a new bond or a strengthening bond is filed pursuant to §201.191 or §201.212.

(72 Stat. 1349; 26 U.S.C. 5171, 5172)

§201.135 Powers of attorney.

The proprietor shall execute and file with the assistant regional commissioner a Form 1534, in accordance with the instructions on the form, for every person authorized to sign or to act on behalf of the proprietor. (Not required for persons whose authority is furnished in the application for registration.)

(72 Stat. 1349; 26 U.S.C. 5172)

§201.136 Operating permits.

Except as provided in §201.138, every person required to file an application for registration under §201.131 shall make application for and obtain an operating permit before commencing any of the following operations:

(a) Distilling for industrial use.

(b) Bonded warehousing of spirits for industrial use.

(c) Denaturing spirits.

(d) Bonded warehousing of spirits (without bottling) for nonindustrial use.

(e) Bottling or packaging of spirits for industrial use.

(f) Any other distilling, warehousing, or bottling operation not required to be covered by a basic permit under the Federal Alcohol Administration Act (49 Stat. 978; 27 U.S.C. 203, 204). Application for such operating permit shall be made on Form 2603 to the assistant regional commissioner.

(72 Stat. 1349, 1370; 26 U.S.C. 5171, 5271)

§201.137 Data for application for operating permits.

Each application on Form 2603 shall be executed under the penalties of perjury, and all written statements, affidavits, and other documents submitted in support of the application shall be deemed to be a part thereof. Applications on Form 2603 shall be prepared in accordance with the headings on the form, and instructions thereon and issued in respect thereto, and shall include the following:

(a) Name and principal business address of the applicant.

(b) Plant address, if different from the business address.

(c) Description of the operation to be conducted for which an operating permit must be obtained.

(d) Statement of type of business organization and of the persons interested in the business, supported by the items of information listed in §201.148.

(e) Trade names (see §201.146).

(f) On specific request of the assistant regional commissioner,

furnish a statement showing whether the applicant or any of the persons whose names and addresses are required to be furnished under the provisions of §201.148 (a)(8) and (c) has—(1) ever been convicted of a felony or misdemeanor under Federal or State law, (2) ever been arrested or charged with any violation of State or Federal law (convictions or arrests on charges for traffic violations need not be reported as to subparagraphs (1) and (2) of this paragraph, if such violations are not felonies), or (3) ever applied for, held, or been connected with a permit, issued under Federal law to manufacture, distribute, sell, or use spirits or products containing spirits, whether or not for beverage use, or held any financial interest in any business covered by any such permit, and, if so, give the number of classification of such permit, the period of operation thereunder, and state in detail whether such permit was ever suspended, revoked, annulled, or otherwise terminated.

Where any of the information required by paragraph (d) of this section is on file with the assistant regional commissioner, the applicant may, by incorporation by reference thereto, state that such information is made a part of the application for an operating permit. The applicant shall, when so required by the assistant regional commissioner, furnish as a part of his application for an operating permit such additional information as may be necessary for the assistant regional commissioner to determine whether the applicant is entitled to the permit.

(72 Stat. 1349, 1370; 26 U.S.C. 5171, 5271)

§201.138 Exceptions to operating permit requirements.

The provisions of §201.136 shall not apply to any agency of a State or political subdivision thereof, or to any officer or employee of any such agency acting for such agency.

(72 Stat. 1349, 1370; 26 U.S.C. 5171, 5271)

§201.147 Major equipment.

The following items of major

equipment, if on the plant premises, shall be described in the application for registration:

(a) Mash tubs and cookers (serial number and capacity).

(b) Fermenters (serial number and capacity).

(c) Tanks used in the production, storage, denaturation, rectification, bottling, and measurement of spirits and tanks used in the storage and the measurement of denatured spirits (designated use (or uses), serial number, capacity, and method of gauging or measurement).

(d) Permanently installed scales and other measuring equipment (including meters).

(e) Bottling lines (list separately as to use and serial number).

(f) Stills (serial number, kind, capacity, and intended use). (The capacity shall be stated as the estimated maximum proof gallons of spirits capable of being produced every 24 hours, or (for column stills) may be represented by a statement of the diameter of the base and number of plates.)

(g) Other items of fixed equipment used in the production, storage, rectification and/or bottling of spirits, if valued at $5,000 or more (description and use).

The description shall show, as to each item of equipment, the location thereof in the plant, and the premises (bonded or bottling) and the facility (production, storage, denaturation, or bottling on bonded premises, and rectification or bottling on bottling premises) in which it is to be used. Where any equipment is to be used in two or more facilities, it shall be identified as for multiple use, and its use in each facility shall be shown.

(72 Stat. 1349; 26 U.S.C. 5172)

§201.148 Organizational documents.

The supporting information required by paragraph (c) of §201.132, and paragraph (d) of §201.137, includes, as applicable:

(a) *Corporate documents.* (1) Certified true copy of articles of incorporation and any amendments thereto.

(2) Certified true copy of the corporate charter or a certificate of corporate existence or incorporation.

(3) Certified true copy of certificate authorizing the corporation to operate in the State where the plant is located (if other than that in which incorporated).

(4) Certified list of directors and officers, showing their names and addresses.

(5) Certified true copy of bylaws.

(6) Certified extracts or digests of minutes of meetings of board of directors, authorizing certain individuals to sign for the corporation.

(7) Statement showing the number of shares of each class of stock or other evidence of ownership, authorized and outstanding, the par value thereof, and the voting rights of the respective owners or holders.

(b) *Articles of partnership.* True copy of the articles of partnership or association, if any, or certificate of partnership or association where required to be filed by any State, county, or municipality.

(c) *Statement of interest.* (1) Names and addresses of the 10 persons having the largest ownership or other interest in each of the classes of stock in the corporation, or other legal entity, and the nature and amount of the stockholding or other interest of each, whether such interest appears in the name of the interested party or in the name of another for him. If a corporation is wholly owned or controlled by another corporation, those persons of the parent corporation who meet the above standards are considered to be the persons interested in the business of the subsidiary, and the names thereof need be furnished to the assistant regional commissioner only at his request.

(2) In the case of an individual owner or partnership, name and address of every person interested in the plant, whether such interest appears in the name of the interested party or in the name of another for him.

(72 Stat. 1349, 1370; 26 U.S.C. 5172, 5271)
[25 FR 6053, June 30, 1960. Redesignated at 40 FR 16835, Apr. 15, 1975, and amended by

T.D. ATF-29, 41 FR 36492; Aug. 30, 1976]

§201.149 Description of plant.

The application for registration shall include a description of each tract of land comprising the plant, clearly indicating the bonded premises, the bottling premises, and any other premises to be included as part of the plant. In the case of a plant producing spirits, where the premises subject to tax lien under section 5004(b), I.R.C., are not coextensive with the bonded premises, the tract of land on which any building containing any part of the bonded premises is situated shall also be described. The description of each tract of land subject to lien under section 5004(b), I.R.C., shall be by courses and distances, in feet and inches (or hundredths of feet), with the particularity required in conveyances of real estate. If any area (or areas) of the plant is to be alternated between bonded and bottling premises, as provided in §201.175, each such area shall be described, and shall be identified by number or letter. The description of denaturing facilities (and equipment) shall show the manner of segregation of such facilities from other facilities which prevents contamination of undenatured spirits. Each building and outside tank shall be described (location, size, construction, arrangement, and means of protection and security), referring to each by its designated number or letter, and use. If a plant consists of a room or floor of a building, a description of the building in which the room or floor is situated and its location therein shall be given.

(72 Stat. 1349; 26 U.S.C. 5172)

§201.150 Registry of stills.

The provisions of Part 196 of this chapter are applicable to stills located on plant premises. The listing of stills for distilling in the application for registration, and the approval of the application for registration, shall constitute registration of such stills.

(72 Stat. 1349, 1355; 26 U.S.C. 5172, 5179)

§201.151 Statement of title.

The application for registration shall include a statement setting forth the name and address of the owner in fee of the lot or tract of land subject to lien under section 5004(b)(1), I.R.C., the buildings thereon, and the equipment used for the production of spirits. If the applicant is not the owner in fee of such property, or if such property is encumbered by mortgage or other lien, the application for registration shall be accompanied by a consent on Form 1602, as provided in §201.152, unless indemnity bond on Form 3A is filed, as provided in §201.200.

(72 Stat. 1349; 26 U.S.C. 5172)

§201.152 Consent on Form 1602.

Consents on Form 1602, where required by this subpart, shall be executed by the owner (if other than the proprietor) of property subject to lien under section 5004(b)(1), I.R.C., and by any mortgagee, judgment creditor, or other person having a lien on such property, duly acknowledging that the property may be used for the purpose of distilling spirits, subject to the provisions of law, and expressly stipulating that the lien of the United States, for taxes on distilled spirits produced thereon and penalties relating thereto, shall have priority of such mortgage, judgment, or other encumbrance, and that in the case of the forfeiture of such property, or any part thereof, the title to the same shall vest in the United States, discharged from such mortgage, judgment, or other encumbrance.

(72 Stat. 1349; 26 U.S.C. 5172, 5173)

§201.153 Statement of process.

The statement of process in the application for registration shall set forth a step-by-step description of the process employed to produce spirits, commencing with the treating, mashing, or fermenting of the raw materials or substances and continuing through each step of the distilling, redistilling, purifying and refining processes to the production gauge, and showing the kind and approximate quantity of each material or substance used in producing, purifying, or refining each type of spirits.

(72 Stat. 1349; 26 U.S.C. 5172)

PLAT AND PLANS

§201.154 General requirements.

The proprietor shall submit, as part of his application for registration, a plat of the premises and plans, in triplicate, as required by this subpart.

(72 Stat. 1349; 26 U.S.C. 5172)

§201.155 Preparation.

Each plat and floor plan shall be drawn to a scale of not less than 1/100 inch per foot and shall show the cardinal points of the compass. Each sheet of the drawings shall—

(a) Bear a distinctive title;

(b) Be numbered in consecutive order, the first sheet being designated number 1; and

(c) Have a clear margin of not less than 1 inch on each side and have outside measurements of 15 by 20 inches: *Provided*, That the assistant regional commissioner may authorize the use of larger sheets if they can be satisfactorily filed.

Plats and plans shall be submitted on tracing cloth, sensitized linen, or blueprint paper, and may be original drawings, or, if clear and distinct, reproductions made by lithoprint, ditto, or ozalid processes. The director may approve other materials and methods which he finds are equally acceptable.

(72 Stat. 1349; 26 U.S.C. 5172)

§201.156 Depiction of plant.

The plat shall show the boundaries of the plant, and delineate separately the portions thereof comprising the bonded premises, the bottling premises, and any other premises to be included as a part of the plant, in feet and inches (or hundredths of feet). The delineation of these premises shall agree with the description given in the application for registration.

The plat shall show (a) all buildings on the plant premises, (b) all basic equipment (including tanks and stills) not located in buildings, and (c) all driveways, public thoroughfares, and railroad rights-of-way contiguous to, connecting, or separating the plant premises. Each building, enclosed area, and outside tank shall be identified. Each pipeline for the conveyance of spirits to and from the premises of the plant, and between bonded and bottling premises, shall be shown on the plat in blue, and each pipeline for the conveyance of denatured spirits to and from the premises of the plant shall be shown on the plat in green: *Provided*, That in lieu of such colors, the pipelines may be identified by symbols which permit ready identification of their uses. The purpose for which such pipelines are used and the points of origin and termination shall be indicated on the plat. Where premises on which spirits, wines, or beer are manufactured, stored, or sold are contiguous to a plant, the plat shall show the relative location of the plant and such contiguous premises, and all pipelines and other connections between them (public utility pipelines and similar connections excepted). The outline of such contiguous premises and of the plant shall be shown in contrasting colors. Where a plant consists of less than an entire building, the plat shall show the building, and the land on which such building is situated. Where a plant consists of, or includes, one or more floors or rooms of a building that is not wholly included in the plant, the floors or rooms so used shall be shown on a floor plan. Each floor plan shall show the location and dimensions of the floors or rooms, the means of ingress and egress, and, insofar as required on plats by this section, pipelines and contiguous premises. Where construction of floors or rooms is identical, a typical plan of such floors or rooms will be acceptable. Where the floor plan shows the entire plant and includes all the information required by a plat, such plan may be accepted in lieu of a plat.

(72 Stat. 1349; 26 U.S.C. 5172)

§201.157 Flow diagrams.

Flow diagrams (plans) shall be submitted reflecting the production processes on bonded premises. The flow diagram shall show major equipment (identified as to use) in its relative operating sequence, with essential connecting pipelines (appropriately identified by color) and valves. The flow diagram shall include the entire closed distilling system. Minor equipment (such as pumps, pressure regulators, rotometers) need not be shown. The direction of flow through the pipelines shall be indicated by arrows.

(72 Stat. 1349; 26 U.S.C. 5172)

§201.158 Certificate of accuracy.

The plat and plans shall bear a certificate of accuracy in the lower right-hand corner of each sheet, signed by the proprietor, substantially as follows:

(Name of proprietor)
(Distilled spirits plant No.)
(Address)

Accuracy certified by:

(Name and capacity— for the proprietor)

Sheet No. _____   Date _____

(72 Stat. 1349; 26 U.S.C. 5172)

Subpart G—Bonds and Consents of Surety

§201.191 General.

Every person intending to commence or to continue the business of a distiller, bonded warehouseman, or rectifier, shall file bond, Form 2601, as prescribed in this subpart, with the assistant regional commissioner, at the time of filing the original application for registration of his plant, and at such other times as are required by this part. Such bond shall be conditioned that he shall faithfully comply with all provisions of law and regulations relating to the duties and business of a distiller, bonded warehouseman, or rectifier, as the case may be (including the payment of taxes imposed by chapter 51 I.R.C.), and shall pay all penalties incurred or fines imposed on him for violation of any such provisions. The assistant regional commissioner may require, in connection with any bond on Form 2601, a statement, executed under the penalties of perjury, as to whether the principal or any person owning, controlling, or actively participating in the management of the business of the principal has been convicted of or has compromised any offense set forth in §201.198(a) or has been convicted of any offense set forth in §201.198(b). In the event the above statement contains an affirmative answer, the applicate shall submit a statement describing in detail the circumstances surrounding such conviction or compromise. Once every four years, and as provided in §201.213, a new bond, Form 2601, shall be executed and filed in accordance with the provisions of this subpart. No person shall commence or continue the business of a distiller, bonded warehouseman, or rectifier, unless he has a valid bond. Form 2601 (and consent of surety, if necessary), as required in respect of such business by this part.

(72 Stat. 1349, 1394; 26 U.S.C. 5173, 5551)

§201.200 Indemnity bond, Form 3A.

A proprietor of a plant qualified for the production of spirits may furnish bond on Form 3A to stand in lieu of future liens imposed under section 5004(b)(1), I.R.C., and no lien shall attach to any lot or tract of land, distillery, building, or distilling apparatus by reason of distilling done during any period included within the term of any such bond. Where an indemnity bond has been furnished on Form 3A in respect of a plant, the requirements of this part relating to the filing of consents on Forms 1602 and bonds on Forms 1617 and Forms 4737 are not applicable in respect to such plant.

(72 Stat. 1317, 1349, as amended; 26 U.S.C. 5004, 5173)

[T.D. 7112, 36 FR 8572, May 8, 1971]

§201.211 Bonds and penal sums of bonds.

The bonds, and the penal sums thereof, required by this subpart, are as follows:

Bond	Penal sum Basis	Min.	Max.
(a) Distiller's, Form 2601	The amount of tax on spirits produced in his distillery during a period of 15 days.	$5,000	$100,000
(b) Bonded Warehouseman's, Form 2601:			
(1) General	The amount of tax on spirits (including denatured spirits) stored on such premises and in transit thereto.	5,000	200,000
(2) Limited to storage of not over 500 wooden packages, and to a total of not over 50,000 proof gallons.	do	5,000	50,000
(3) Limited to storage of denatured spirits, denaturation of spirits, and storage of not to exceed 100,000 proof gallons of spirits prior to denaturation.	do	5,000	100,000
(c) Rectifier's, Form 2601	The amount of tax the rectifier will be liable to pay in a period of 30 days under sections 5021 and 5022, I.R.C.	1,000	100,000
(d) Combined Operations, Form 2601:			
(1) Distiller and bonded warehouseman	Sum of penal sums of bonds in lieu of which given.	10,000	200,000
(2) Distiller and rectifier	do	6,000	200,000
(3) Bonded, warehouseman and rectifier.	do	6,000	250,000

(continued next page)

	Penal sum		
	Basis	Min.	Max.
(4) Distiller bonded warehouseman, and rectifier.	do	11,000	250,000
(5) Distiller and bonded wine cellar.	do	6,000	150,000
(6) Distiller, bonded warehouseman, and bonded wine cellar.	do	.11,000	250,000
(7) Distiller, rectifier, and bonded wine cellar.	do	7,000	250,000
(8) Distiller, bonded warehouseman, rectifier, and bonded wine cellar.	do	12,000	300,000
(e) Blanket bond, Form 2601	The penal sum shall be calculated in accordance with the following table:		

Total penal sums as determined under (a), (b), (c), and (d).	Requirements for penal sum of blanket bond.
Not over $300,000	100 percent.
Over $300,000 but not over $600,000.	$300,000 plus 70 percent of excess over $300,000.
Over $600,000 but not over $1,000,000.	$510,000 plus 50 percent of excess over $600,000.
Over $1,000,000 but not over $2,000,000.	$710,000 plus 35 percent of excess over $1,000,000
Over $2,000,000	$1,060,000 plus 25 percent of excess over $2,000,000

	Basis	Min.	Max.
(f) Indemnity, Form 3A	Appraised value of property		$300,000
(g) Indemnity, Form 1617	Decrease in value of property.	$5,000	300,000
(h) Indemnity Bond, Form 4737	The amount of involuntary liens against property.	(1)	(1)
(i) Withdrawal Bond, Form 2613.	The amount of tax which, at any one time, is chargeable against such bond but has not been paid.	1,000	1,000,000
(j) Withdrawal Bond, Form 2614.	do	1,000	1,000,000
(k) Blanket Withdrawal Bond Form 2615:			
(1) Bonded and bottling premises on same plant premises.	Sum of penal sums of bonds, Forms 2613 and 2614, in lieu of which given.	2,000	1,000,000
(2) Two or more plants in a region qualified for bonded and/or bottling operations.	Sums of the penal sums of all the bonds, Forms 2613 and/or 2614, in lieu, of which given.	(2)	(3)

[1] Sum of outstanding involuntary lien or liens covered by the bond.
[2] Sum of the minimum penal sums required for each plant covered by the bond.
[3] Sum of the maximum penal sums required for each plant covered by the bond. (The maximum penal sum for one plant is $1,000,000.)

(68A Stat. 847, 72 Stat. 1349, as amended, 1352; 26 U.S.C. 7102, 5173, 5174, 5175)

[25 FR 6053, June 30, 1960, as amended by T.D. 7112, 36 FR 8572, May 8, 1971]

Subpart H—Construction and Equipment

§201.231 Protection of premises.

(a) *Buildings.* The buildings in which spirits are held or stored shall be securely constructed of masonry, concrete, wood, metal, or other equally substantial material, and arranged, equipped, and protected to afford adequate supervision and inspection by internal revenue officers. Except for doors or other openings approved by the assistant regional commissioner, separations between bonded premises and bottling premises shall be secure and unbroken.

(b) *Doors and windows.* Doors to rooms or buildings under the joint custody of the assigned officer and the proprietor shall be so installed and equipped as to prevent their removal and shall be rigidly secure when locked. Windows in such rooms or buildings shall be equipped with sash locks or comparable fasteners. If the location of such windows with respect to the ground, to a fire escape, roof, set back, or balcony, or to an adjacent or contiguous structure would permit ingress to such rooms or buildings or would otherwise, in the opinion of the assistant regional commissioner, create a jeopardy to security, the assistant regional commissioner shall require such windows to be of the detention type equipped with wire glass panes or to be protected by means of iron bars or shutters or other means affording equal protection to the revenue.

(c) *Other openings.* Skylights, monitors, penthouses, and similar roof openings in such rooms or buildings shall be treated as windows for security purposes. Ventilating or heating ducts, or sewerage or drainage openings which would permit ingress to such rooms or buildings, or would otherwise, in the opinion of the assistant regional commissioner, create a jeopardy to the revenue, shall be protected with secure metal grills or other means which the assistant regional commissioner considers to be equally effective.

(d) *Additional security.* Where the assistant regional commissioner finds the construction, arrangement, equipment, or protection inadequate, he shall require additional security to be provided (such as fences, floodlights, alarm systems, watchman services) or changes in construction, arrangement, or equipment to be made to the extent necessary to assure him that the construction, arrangement, or equipment is adequate to protect the revenue.

§201.234 Locking of storage rooms or buildings on bonded premises.

Where spirits are bottled or packaged, or stored in packages or in cases or in other portable containers on bonded premises, the proprietor shall provide a room or building for such bottling, packaging, or storage. Such room or building shall be constructed as provided in §201.231, arranged and equipped so as to be suitable for the intended purpose, and shall be equipped for locking with Government locks. Any other building, room, or enclosure on bonded premises, not secured by Government lock, in which spirits (including denatured spirits) are held, shall be equipped for locking by the proprietor.

§201.236 Identification of structures, areas, apparatus, and equipment.

Each room or enclosed area where spirits (including denatured spirits) or wines, distilling or fermenting materials, or containers are held, and each building, within the plant, shall be appropriately marked as to use. Each tank or receptacle for spirits (including denatured spirits) or wine shall be marked to show its serial number, capacity, and use. Where tanks or receptacles are used for multiple purposes, such uses shall be indicated. Each still shall be numbered and marked to show its use. All other major equipment used for processing or containing spirits (including denatured spirits) or wine, or distilling or fermenting material, and all other tanks, shall be identified as to use unless the intended use thereof is readily apparent.

(72 Stat. 1353; 26 U.S.C. 5178)

§201.238 Government office.

The proprietor shall provide an office at the plant for the exclusive use of Government officers in performing supervisory and administrative duties and safeguarding Government records and property. Such office shall be adequately equipped with office furniture and a secure cabinet fitted for locking with a Government lock and with toilet and lavatory facilities, shall be well lighted, ventilated, and heated, and shall be subject to the approval of the assistant regional commissioner. Where suitable facilities are otherwise available, the assistant regional commissioner may waive the requirements for a separate Government office.

§201.240 Closed distilling system.

The distilling system shall be continuous and closed at all points where potable or readily recoverable spirits are present, and shall be so designed, constructed, and connected as to prevent the unauthorized removal therefrom without detection, of such spirits and of distilling material and stillage from which spirits are readily recoverable. The security of the closed distilling system shall be maintained, and removal of spirits therefrom controlled by Government locks or seals, or by such meters or other devices or methods affording comparable protection as may be approved by the Director. Processing equipment not susceptible of being locked or sealed (a) shall be located in a room or enclosure which shall be in the joint custody of the assigned officer and the proprietor, which room or enclosure shall be locked with a Government lock, and shall not be unlocked or remain unlocked except when such officer is on the plant premises, or (b) shall be otherwise equally protected in a manner approved by the Director. In addition, processing equipment not located within a room or building shall, unless the premises or general area containing such equipment is enclosed within a fence or wall which the assistant regional commissioner deems adequate to the protection of the revenue, be enclosed within a secure fence constructed as provided in §201.243(b). The provisions of this section do not preclude the removal of spirits from the closed distilling system, pursuant to production gauge, and their deposit in bonded storage, by redesignation of the tank in which the production gauge is made as a storage tank, and the necessary locking or sealing of the tank to remove it from the closed system.

(72 Stat. 1353; 26 U.S.C. 5203)

§201.242 Denaturing facilities.

Where the proprietor is authorized to denature spirits, he shall provide on his bonded premises, segregated facilities for use in his denaturing operations. These facilities shall include:

(a) A storage room or rooms, fitted for locking, for storing packages of denaturants, if denaturants are to be received in packages;

(b) Tanks and other suitable receptacles, fitted for locking, for storing denaturants, if denaturants are to be received in bulk quantities;

(c) Tanks, fitted for locking, for denaturing spirits, if spirits are to be denatured in tanks;

(d) Tanks for storing denatured spirits, if denatured spirits are to be stored on bonded premises in tanks; and

(e) Storage space in a room or building for storing packages of denatured spirits, if denatured spirits are to be removed in packages.

The proprietor shall also provide such meters and other equipment and apparatus as may be necessary to carry out proposed operations. Meters, equipment, and facilities used for handling or processing denaturing materials or denatured spirits shall not be used for spirits other than spirits to be denatured: *Provided,* That the assistant regional commissioner may authorize their use for other spirits when he determines that contamination of such other spirits will not take place.

(72 Stat. 1353; 26 U.S.C. 5178)

SPECIFIC REGULATORY REQUIREMENTS

Applicable Excerpts From
27 CFR Part 211—
Distribution and Use of
Denatured Alcohol and Rum

**Subpart D—Qualification of Bonded Dealers
and Users**

APPLICATION FOR INDUSTRIAL USE
PERMIT

§211.42 Application, Form 1479, for permit to use or recover.

Every person desiring to use specially denatured alcohol or specially denatured rum, or both, and every person desiring to recover denatured alcohol, specially denatured rum, or articles shall, before commencing business, make application for and obtain an industrial use permit, Form 1481. Except as provided in §211.42a, application for an industrial use permit shall be on Form 1479. Such application, and necessary supporting documents as required by this subpart for such permit, shall be filed with the assistant regional commissioner. All data, written statements, affidavits, and other documents submitted in support of the application shall be deemed to be a part thereof. Such application shall be accompanied by evidence which will establish the authority of the officer or other person who executes the application to execute the same and, where applicable, by the application for a withdrawal permit, Form 1485, required by §211.161.

(72 Stat. 1370; 26 U.S.C. 5271)
[T.D. 7058, 35 FR 14395, Sept. 12, 1970]

§211.43 Data for application, Forms 1474* and 1479.

Each application on Form 1474* or 1479 shall include, as applicable, the following information:

(a) Serial number and purpose for which filed.

(b) Name and principal business address of applicant.

(c) Location of the dealer's or user's premises if different from the business address.

(d) Statement as to the type of business organization and of the persons interested in the business, supported by the items of information listed in §211.53.

(e) Statement of operations showing the estimated maximum quantity in gallons of specially denatured alcohol or specially denatured rum to be on hand, in transit, and unaccounted for at any one time and, in the case of users, a general statement as to the intended use to be made of the specially denatured alcohol or specially denatured rum, and whether recovery, restoration, and redenaturation processes will be used, and, if so, the estimated number of gallons of recovered denatured alcohol, recovered specially denatured rum, or recovered articles to be on hand at any one time.

(f) Listing of principal equipment to be used in manufacturing, packaging, and recovery processes, including processing tanks, storage tanks, bottling facilities, and equipment for the recovery, restoration (including the serial number, kind, capacity, name and address of owner, and intended use of distilling apparatus), and redenaturation of recovered denatured alcohol or specially denatured rum by users, and the size and complete description of the specially denatured alcohol or specially denatured rum storeroom or storage tanks.

(g) Trade names (see §211.52).

(h) List of the offices, the incumbents of which are authorized by the articles of incorporation, by laws, or the board of directors to act on behalf of the applicant or to sign his name.

(i) On specific request of the assistant regional commissioner, furnish a statement showing whether any of the persons whose names and addresses are required to be furnished under the provisions of §§211.53(a)(2) and 211.53(c) have (1) ever been convicted of a felony or misdemeanor under Federal or State law, (2) ever been arrested or charged with any violation of State or Federal law (convictions or arrests or charges for traffic violations need not be reported as to subparagraphs (1) and (2) of this paragraph, if such violations are not felonies), or (3) ever applied for, held, or been connected with a permit issued under Federal law to manufacture, distribute, sell, or use spirits or products containing alcohol or rum, whether or not for beverage use, or held any financial interest in any business covered by any such permit, and, if so, give the number and classification of such permit, the period of operation thereunder, and state in detail whether such permit was ever suspended, revoked, annulled, or otherwise terminated.

Where any of the information required by paragraphs (d) through (h) of this section is on file with the assistant regional commissioner, the applicant may, by incorporation by reference thereto, state that such information is made a part of the application for an industrial use permit. The applicant shall, when so required by the assistant regional commissioner, furnish as part of his application for an industrial use permit such additional information as may be necessary for the assistant regional commissioner to determine whether the applicant is entitled to the permit.

(72 Stat. 1370; 26 U.S.C. 5271)

*Bonded dealer only

§211.53 Organizational documents.

The supporting information required by paragraph (d) of §211.43 includes, as applicable:

(a) *Corporate documents.* (1) Certified true copy of the certificate of incorporation, or certified true copy of certificate authorizing the corporation to operate in the State where the premises are located (if other than that in which incorporated).

(2) Certified list of names and addresses of officers and directors.

(3) Statement showing the number of shares of each class of stock or other evidence of ownership, authorized and outstanding, the par value thereof, and the voting rights of the respective owners or holders.

(b) *Articles of partnership.* True copy of the articles of partnership or association, if any, or certificate of partnership or association where required to be filed by any State, county, or municipality.

(c) *Statement of interest.* (1) Names and addresses of the 10 persons having the largest ownership or other interest in each of the classes of stock in the corporation, or other legal entity, and the nature and amount of the stockholding or other interest of each, whether such interest appears in the name of the interested party or in the name of another for him. If a corporation is wholly owned or controlled by another corporation, those persons of the parent corporation who meet the above standards are considered to be the persons interested in the business of the subsidiary and the names and addresses of such persons shall be submitted to the assistant regional commissioner on his specific request.

(2) In the case of an individual owner or partnership, name and address of every person interested in the business, whether such interest appears in the name of the interested party or in the name of another for him.

§211.54 Powers of attorney.

An applicant or permittee shall execute and file with the assistant regional commissioner a Form 1534, in accordance with the instructions on the form, for every person authorized to sign or to act on his behalf. (Not required for persons whose authority is furnished in accordance with §211.43 or §211.43a.)

[T.D. 7058, 35 FR 14396, Sept. 12, 1970]

§211.72 User's bond, Form 1480.

Every person filing an application on Form 1479 shall, before issuance of the industrial use permit, file bond, Form 1480, with the assistant regional commissioner, except that no bond will be required where the application is filed by a State, or any political subdivision thereof, or the District of Columbia, or where the quantity of specially denatured alcohol and specially denatured rum authorized to be withdrawn does not exceed 120 gallons per annum and the quantity which may be on hand, in transit, and unaccounted for at any one time does not exceed 12 gallons. The penal sum of the bond shall be computed on each gallon of specially denatured alcohol or rum, including recovered or restored denatured alcohol or specially denatured rum or recovered articles in the form of denatured spirits, authorized to be on hand, in transit to the premises of the user, and unaccounted for at any one time, at double the rate prescribed by law as the internal revenue tax on a proof gallon of distilled spirits: *Provided,* That the penal sums of bonds covering specially denatured alcohol Formulas No. 18 and No. 19 shall be computed on each gallon at the rate prescribed by law as the tax on a proof gallon of distilled spirits. The penal sum of any such bond (or the total of the penal sums where original and strengthening bonds are filed) shall not exceed $100,000 or be less than $500. No bond is required where application is filed on Form 4326, as provided in §211.42A.

(72 Stat. 1372; 26 U.S.C. 5272)
[T.D. 7058, 35 FR 14396, Sept. 12, 1970]

§211.91 Premises

A permittee shall have premises suitable for the business being conducted and adequate for the protection of the revenue. When specially denatured spirits are to be stored, storage facilities shall be provided on the premises for such spirits received or recovered thereon. Except as otherwise provided in this section, these storage facilities shall consist of storerooms or stationary storage tanks (not necessarily in a room or building), or a combination thereof. A user receiving specially denatured spirits in tank cars or tank trucks and storing all such spirits therein, as provided in §211.168, need not provide stationary storage tanks. Where specially denatured spirits are to be received at or removed from a permittee's premises in bulk conveyances, suitable facilities for such operations shall be provided.

(72 Stat. 1372; 26 U.S.C. 5273)

§211.101 General.

(a) *Form 1479-A.* Every person desiring to use specially denatured spirits for other than laboratory or mechanical purposes, as provided in §211.169, or to recover denatured spirits or articles, shall, except where previously approved formulas are adopted or as provided in §211.102, submit on Form 1479-A, directly to the Director, a description of each process or formula; a separate Form 1479-A shall be used for each such formula or process. In the case of articles to be manufactured with specially denatured spirits, quantitative formulas and processes shall be given. The preparation of Form 1479-A shall be in accordance with the headings and the instructions thereon.

(b) *Previously approved Forms 1479-A.* Any persons who intends to use previously approved formulas and processes, Forms 1479-A, on and after July 1, 1960, shall submit to the assistant regional commissioner a list, in quadruplicate, of all such approved Forms 1479-A which he intends to continue using. The list shall show, as to each Form 1479-A, the article or process in which denatured spirits are used or recovered, the formula of specially denatured spirits, the laboratory number of the sample (if any), the date of approval, and the code number prescribed for the article or process.

(72 Stat. 1369, 1372; 26 U.S.C. 5241, 5273)
[25 FR 5968, June 28, 1960, as amended by T.D. 7058, 35 FR 14397, Sept. 12, 1970]

§211.107 Samples of articles and ingredients.

In connection with the submission of Form 1479-A covering the proposed manufacture of an article (except a rubbing alcohol, a rubbing alcohol base, a proprietary solvent, or a special industrial solvent) containing specially denatured spirits, the applicant shall submit to the Director an 8-ounce sample of the article (except that a 4-ounce sample will be sufficient for a perfume which contains more than 6 ounces of perfume oils per gallon). For all toilet preparations containing specially denatured spirits, the applicant shall also submit a 1-ounce sample of the perfume oils (or of purchased mixtures consisting of perfume oils with other ingredients) to be used. The Director may at any time require the submission of samples of (a) any ingredients included in a formula, and (b) proprietary antifreeze solutions containing completely denatured alcohol.

(72 Stat. 1372; 26 U.S.C. 5273)
[37 FR 5751, Mar. 21, 1972]

Subpart H—Sale and Use of Completely Denatured Alcohol

§211.111 General

Completely denatured alcohol may be sold and used for any lawful purpose. Completely denatured alcohol may be used (a) in the manufacture of definite chemical substances where such alcohol is changed into some other chemical substance and does not appear in the finished product; (b) in the arts and industries (except in the manufacture of preparations or products for internal human use or consumption where any of such alcohol or of the denaturants used in such alcohol may remain in the finished product); and (c) for fuel, light, and power. Use of completely denatured alcohol in the arts and industries includes, but is not limited to, the manufacture of cleaning fluids, detergents, proprietary antifreeze solutions, thinners, lacquers, and brake fluids. Persons distributing and using (but not recovering for reuse) completely denatured alcohol are not required to obtain a permit or to file bond under this part. Persons recovering completely denatured alcohol for reuse shall procure an industrial use permit in accordance with Subpart D of this part and file bond in accordance with Subpart E of this part. Containers of products manufactued with completely denatured alcohol, such as proprietary antifreeze preparations, solvents, thinners, and lacquers, shall not be branded as completely denatured alcohol nor shall any such product be advertised, shipped, sold, or offered for sale as completely denatured alcohol.

(72 Stat. 1362, 1369, 1372; U.S.C. 5214, 5241, 5273)

Subpart J—Operations by Users of Specially Denatured Spirits

PROPRIETARY SOLVENTS

§211.170 Manufacture of proprietary solvents.

All articles coming under the general classification of proprietary solvents shall be manufactured with specially denatured alcohol Formula No. 1. The formulations shall be as follows, except as may otherwise be authorized by the Director:

(1) Formulation No. I.	Gallons
Specially denatured alcohol formula No. 1	100
Ethyl acetate	5
Gasoline or rubber hydrocarbon solvent.	
(2) Formulation No. II.	
Specially denatured alcohol formula No. 1	100

Denaturing grade wood alcohol	2
Ethyl acetate	1
Gasoline or rubber hydrocarbon carbon	1
(3) Formulation No. III.	
Specially denatured alcohol formula No. 1	100
Methyl isobutyl ketone	1
Ethyl acetate	1
Gasoline or rubber hydrocarbon solvent.	1
(4) Formulation No. IV.	
Specially denatured alcohol formula No. 1	100
Methyl isobutyl ketone	1
tert-butyl alcohol	2
Gasoline or rubber hydrocarbon solvent.	1
(5) Formulation No. V.	
Specially denatured alcohol formula No. 1	100
Methyl isobutyl ketone	1
Secondary butyl alcohol	2
Gasoline or rubber hydrocarbon solvent	1

§211.171 Sales by producers.

Proprietary solvents may be sold by producers to any person for use in manufacturing or as a solvent and to distributors and other persons for resale.

§211.172 Use in manufacturing.

When a proprietary solvent is used in the manufacture, for sale, of an article containing more than 25 percent of alcohol by volume, sufficient ingredients shall be added to definitely change the composition and character of the proprietary solvent. Such articles shall not be manufactured until a Form 1479-A covering production of the article has been submitted to and approved by the Director, except that Form 1479-A need not be submitted to cover the manufacture of surface coatings (including such products as inks) containing two pounds or more of solid coating material per gallon of such article. The formulation number (see §211.170) of the proprietary solvent to be used in manufacturing the article shall be stated in the Form 1479-A.

[T.D. 6715, 29 FR 3659, Mar. 24, 1964]

§211.180 Manufacture.

Special industrial solvents shall be manufactured with specially denatured alcohol Formula No. 1 or 3-A. The formulations shall be as follows, except as may otherwise be authorized by the Director:

(1) *Formulation A.*	*Gallons*
Specially denatured alcohol Formula No. 1 or 3-A	100
Isopropyl alcohol or methyl alcohol .	10
Methyl isobutyl ketone	1

(2) *Formulation B.*	
Specially denatured alcohol Formula No. 1 or 3-A	100
Isopropyl alcohol	5
Methyl isobutyl ketone	1
Methyl alcohol	5

(3) *Formulation C.*	
Specially denatured alcohol Formula No. 1 or 3-A	100
Methyl isobutyl ketone	1

(4) *Formulation D.*	
Ethyl acetate	5
Specially denatured alcohol Formula No. 1 or 3-A	100
Isopropyl alcohol or methyl alcohol .	15
Methyl isobutyl ketone	1

[T.D. 6715, 29 FR 3659, Mar. 24, 1964]

§211.181 Sales by producers.

Special industrial solvents may be sold by producers to any person for use in manufacturing or as a solvent, and to wholeslae distributors and other producers of such solvents for resale. Sale of such solvents for distribution through retail channels is prohibited.

[T.D. 6715, 29 FR 3659, Mar. 24, 1964]

§211.182 Use in manufacuring articles for sale.

When a special industrial solvent is used in the manufacture of an article for sale, sufficient ingredients shall be added to definitely change the composition and character of the special industrial solvent; such an article shall not be manufactured until a Form 1479-A covering its production has been submitted to, and approved by, the Director. The formulation letter (see §211.180) of the special industrial solvent to be used shall be stated in the Form 1479-A. Special industrial solvents shall not be reprocessed into other solvents intended for sale where the other solvent would contain more than 50 percent alcohol by volume.

[37 FR 5752, Mar. 21, 1972]

SPECIFIC REGULATORY REQUIREMENTS

Applicable Excerpts From
27 CFR Part 212—
Formulas for Denatured
Alcohol and Rum

Subpart C—Completely Denatured Alcohol Formulas

§212.10 General.

Completely denatured alcohol will be denatured in accordance with formulas prescribed in this subpart. Producers of completely denatured alcohol may be authorized to add a small quantity of an odorant, rust inhibitor, or dye to completely denatured alcohol. Any such addition may be made only on approval by the Director. Request for such approval shall be submitted to the Director in triplicate. Odorants or perfume materials may be added to denaturants authorized for completely denatured alcohol in amounts not greater than 1 part to 250, by weight: *Provided,* That such addition shall not decrease the denaturing value nor change the chemical or physical constants beyond the limits of the specifications for these denaturants as prescribed in subpart E, except as to odor. Proprietors of distilled spirits plants using denaturants to which such odorants or perfume materials have been added shall inform the Director of the names and properties of the odorants or perfume materials so used.

[22 FR 1330, Mar. 5, 1957, as amended by T.D. 6474, 25 FR 5988, June 29, 1960]

§212.11 Formula No. 18.

To every 100 gallons of ethyl alcohol of not less than 160° proof add:

2.50 gallons of methyl isobutyl ketone;
0.125 gallon of pyronate or a compound similar thereto;
0.50 gallon of acetaldol (b-hydroxybutyraldehyde); and
1.00 gallon of either kerosene, deodorized kerosene, or gasoline.

[T.D. 6634, 28 FR 1038. Feb. 2, 1963]

§212.12 Formula No. 19

To every 100 gallons of ethyl alcohol of not less than 160° proof add:

4.0 gallons of methyl isobutyl ketone; and
1.0 gallon of either kerosene, deodorized kerosene, or gasoline.

[T.D. 6634, 28 FR 1038. Feb. 2, 1963]

Subpart D—Specially Denatured Spirits Formulas and Authorized Uses

§212.15 General.

(a) *Formulas.* Specially denatured alcohol shall be denatured in accordance with formulas prescribed in this subpart. Alcohol of not less than 185° of proof shall be used in the manufacture of all formulas of specially denatured alcohol, unless otherwise authorized by the Director. Rum for denaturation shall be of not less than 150° of proof and shall be denatured in accordance with Formula No. 4.

(b) *Uses.* Users and manufacturers holding approved Forms 1479-A covering manufacture of products or use in processes no longer authorized for a particular formula may continue such use. Subject to the provisions of Chapter 51 I.R.C., Part 211 of this chapter, and this part, the Director may authorize, in his discretion, the use of any formula of specially denatured alcohol or specially denatured rum for uses not specifically authorized in this part. The code number before each item under "authorized uses" shall be used in reporting the use of specially denatured alcohol or specially denatured rum.

[22 FR 1330, Mar. 5, 1957, as amended by T.D. 6474, 25 FR 5988, June 29, 1960; T.D. 6977, 33 FR 15708, Oct. 24, 1968]

§212.16   Formula No. 1

(a) *Formula.* To every 100 gallons of alcohol add:

Five gallons wood alcohol.

(b) *Authorized uses.* As a fuel:

611. Automobile and supplementary fuels.
612. Airplane and supplementary fuels.
613. Rocket and jet fuels.
620. Proprietary heating fuels.

§212.19   Formula No. 3-A.

(a) *Formula.* To every 100 gallons of alcohol add:

Five gallons methyl alcohol.

(b) *Authorized uses.* As a fuel:

611. Automobile and supplementary fuels.
612. Airplane and supplementary fuels.
613. Rocket and jet fuels.
620. Proprietary heating fuels.
630. Other fuel uses.

§212.38   Formula No. 28-A.

(a) *Formula.* To every 100 gallons of alcohol add:

One gallon of gasoline.

(b) *Authorized uses.* As a fuel:

611. Automobile and supplementary fuels.
612. Airplane and supplementary fuels.
613. Rocket and jet fuels.
620. Proprietary heating fuels.
630. Other fuel uses.

FORMS GENERALLY REQUIRED TO BE PREPARED

For Qualification of a Distilled Spirits Plant

ATF Form 3A	Indemnity Bond
ATF Form 1534 (5000.8)	Power of Attorney
ATF Form 1602	Consent
ATF Form 2601 (5110.56)	Distilled Spirits Bond
ATF Form 2603 (5110.25)	Application for Operating Permit Under 26 U.S.C. 517(b)
ATF Form 2607 (5110.41)	Registration of Distilled Spirits Plant
ATF Form 2625 (5000.9)	Personnel Questionnaire— Alcohol and Tobacco Products
ATF Form 4805 (1740.2)	Supplemental Information on Water Quality Considerations —Under 33 U.S.C. 1341(a)
ATF Form 4871 (1740.1)	Environmental Information
ATF Form 5100.1	Signing Authority for Corporate Officials

FORMS GENERALLY REQUIRED TO BE PREPARED

For Qualification of a Specially Denatured Alcohol User's Premises

ATF Form 1479 (5150.23)	Application for Permit to Use Specially Denatured Alcohol
ATF Form 1479-A (5150.19)	Formula for Article Made with Specially Denatured Alcohol or Rum
ATF Form 1480 (5150.20)	Bond of User of Specially Denatured Alcohol or Rum
ATF Form 1485 (5150.12)	Application and Withdrawal Permit of User to Procure Specially Denatured Alcohol
ATF Form 1534 (5000.8)	Power of Attorney
ATF Form 2625 (5000.9)	Personnel Questionnaire— Alcohol and Tobacco Products
ATF Form 4805 (1740.2)	Supplemental Information on Water Quality Considerations —Under 33 U.S.C. 1341(a)
ATF Form 4871 (1740.1)	Environmental Information
ATF Form 5100.1	Signing Authority for Corporate Officials

BATF REGIONAL OFFICES

Central Region
Indiana, Kentucky
Michigan, Ohio,
West Virginia

Regional Regulatory Administrator
Bureau of Alcohol, Tobacco, and Firearms
550 Main Street
Cincinnati, OH 45202
Phone (513) 684-3334

Mid-Atlantic Region
Delaware, District of
Columbia, Maryland,
New Jersey, Pennsylvania,
Virginia

Regional Regulatory Administrator
Bureau of Alcohol, Tobacco, and Firearms
2 Penn Center Plaza, Room 360
Philadelphia, PA 19102
Phone (215) 597-2248

Midwest Region
Illinois, Iowa, Kansas,
Minnesota, Missouri,
Nebraska, North Dakota,
South Dakota, Wisconsin

Regional Regulatory Administrator
Bureau of Alcohol, Tobacco, and Firearms
230 S. Dearborn Street, 15th Floor
Chicago, IL 60604
Phone (312) 353-3883

North-Atlantic Region
Connecticut, Maine,
Massachusetts, New Hampshire,
New York, Rhode Island,
Vermont, Puerto Rico,
Virgin Islands

Regional Regulatory Administrator
Bureau of Alcohol, Tobacco, and Firearms
6 World Trade Center, 6th Floor
(Mail: P.O. Box 15,
Church Street Station)
New York, NY 10008
Phone (212) 264-1095

Southeast Region
Alabama, Florida, Georgia,
Mississippi, North Carolina,
South Carolina, Tennessee

Regional Regulatory Administrator
Bureau of Alcohol, Tobacco, and Firearms
3835 Northeast Expressway
(Mail: P.O. Box 2994)
Atlanta, GA 30301
Phone (404) 455-2670

Southwest Region
Arkansas, Colorado,
Louisiana, New Mexico,
Oklahoma, Texas, Wyoming

Regional Regulatory Administrator
Bureau of Alcohol, Tobacco, and Firearms
Main Tower, Room 345
1200 Main Street
Dallas, TX 75202
Phone (214) 767-2285

Western Region
Alaska, Arizona,
California, Hawaii,
Idaho, Montana,
Nevada, Oregon,
Utah, Washington

Regional Regulatory Administrator
Bureau of Alcohol, Tobacco, and Firearms
525 Market Street, 34th Floor
San Francisco, CA 94105
Phone (415) 556-0226

SAMPLE APPLICATION FOR EXPERIMENTAL DISTILLED SPIRITS PLANT

The following correspondence is representative of what would be needed in the permit application for an experimental distilled spirits plant.

Department of the Treasury
Bureau of Alcohol, Tobacco & Firearms

Refer To
DS-TI-T:MD
EXD-SDM
Phone: 214-767-2285

Farm Fuel, Inc.
1617 Cole Boulevard
Golden, Colorado 80401

Gentlemen:

Your letter application to establish and operate an experimental distilled spirits plant under the provisions of Title 27, Code of Federal Regulations 201.65 has been approved. This authorization will be effective for a period of two (2) years commencing with the date on our approval of your bond.

For the purposes of this application, we are waiving the provisions of Title 27, Code of Federal Regulations, Part 201, except those sections pertaining to the attachment, assessment, and collection of tax (for unauthorized use of the distilled spirits); security applicable to the experimental plant; and bond requirements. In addition, we have not waived the authorities of BATF Officers, including plant and required records.

For each production run of alcohol, you shall maintain a record showing the date(s) of production, kind(s) and quantity(ies) of raw materials used, and quantity and proof of alcohol produced. In addition, you shall maintain a record of disposition of all alcohol produced showing the dates and manner of disposition. The alcohol shall be stored under lock and access thereto shall be limited to personnel essential to the experimentation. Upon termination of this authorization, any alcohol remaining shall be destroyed under Government supervision.

Alcohol produced by the experimental plant may be used at your experimental distilled spirits premises in pure form, after denaturation, or after mixing with gasoline or other additives. However, any alcohol intended for removal from the plant premises, such as fuel in an automobile to be operated on public roads, must be completely denatured according to a formula prescribed in 27 CPR Part 212 and included in BATF 5000.1. Appropriate records covering the denaturing operations must also be maintained. The records shall show the quantity of alcohol used, kind(s) and quantity(ies) of material added, and the disposition of the resultant mixture. Distilled spirits produced by the experimental plant or fuel mixtures made with such spirits may not be sold or given away.

This authorization does not relieve you of any responsiblity for complying with state and local requirements relating to the production and use of ethyl alcohol and may be rescinded should jeopardy to the revenue or administrative problem arise.

Sincerely yours,

Earl P. Kennard
Regional Regulatory Administrator

DISTILLED SPIRITS BOND
(See instructions on back)

TYPE OF BOND

☒ DISTILLER'S ☐ RECTIFIERS ☐ BLANKET
☐ BONDED WAREHOUSEMAN'S ☐ COMBINED OPERATIONS

PRINCIPAL *(See instructions 2, 3, and 4.)*

Farm Fuel, Inc.   A Colorado Corporation
Dave Strawman, Pres.

ADDRESS OF BUSINESS OFFICE *(Number, street, city, State, ZIP Code)*

1617 Cole Boulevard
Golden, Colorado   80401

SURETY (OR SURETIES)
The Universal Insurance Co. of Colorado, Inc.

AMOUNT OF BOND
30,000.00

EFFECTIVE DATE
1/2/80

KIND OF BOND *(Check applicable box.)*   ☒ ORIGINAL   ☐ STRENGTHENING   ☐ SUPERSEDING

KNOW ALL MEN BY THESE PRESENTS, That we, the above-named principal and surety (or sureties), are held and firmly bound unto the United States of America in the above amount, lawful money of the United States, for the payment of which we bind ourselves, our heirs, executors, administrators, successors, and assigns, jointly and severally, firmly by these presents.

This bond shall not in any case be effective before the above date, but if accepted by the United States it shall be effective according to its terms on and after that date without notice to the obligors. *Provided,* That if no effective date is inserted in the space provided, the date of execution shown below shall be the effective date of the bond.

WHEREAS, every person intending to commence or continue the business of a distiller, or of a bonded warehouseman, or of a rectifier, shall give bond in accordance with the provisions of Title 26 United States Code; and

WHEREAS, under the provisions of Title 26 United States Code, any person intending to commence or continue business as proprietor of a distilled spirits plant who would otherwise be required to give more than one bond to cover his several operations shall, in lieu thereof, give a combined operations bond to cover his operations; and

WHEREAS, under the provisions of Title 26 United States Code, any person intending to commence or continue business as proprietor of a bonded wine cellar and an adjacent distilled spirits plant qualified for the production of distilled spirits shall, in lieu of the bonds which would otherwise be required by law, give a combined operations bond to cover his operations; and

WHEREAS, under the provisions of Title 26 United States Code and regulations, any person (including, in the case of a corporation, controlled or wholly owned subsidiaries) operating more than one distilled spirits plant in an alcohol, tobacco and firearms region may give a blanket bond, in lieu of separate bonds, covering the operation of any two or more such plants and any bonded wine cellars which are adjacent to any such plant qualified for the production of distilled spirits; and

WHEREAS, the principal is operating, or intends to operate, the premises specified—

(List on back under "Description of Premises Operated" the name, address, and registry number of each plant; and the operations for which such plant is qualified. If space is insufficient, attach appropriately identified separate sheet of approximately the same size.)

as described in the (several) application(s) for registration (or approval), and (all of) which plant(s) (is) (are) located in the ___Southwest Region___ alcohol, tobacco and firearms region.

NOW, THEREFORE, the conditions of this bond are such that if the principal—

1. Shall, in all respects, without fraud or evasion, comply with all the provisions of law and regulations relating to the duties and business or businesses for which this bond is given; and

2. Shall pay all penalties incurred and fines imposed on him for violation of any of the said provisions; and

3. Shall:

   (a) as a distiller, pay, or cause to be paid, to the United States, all taxes imposed by law now or hereafter in force, plus penalties if any, and interest, with respect to the business of a distiller, and on all distilled spirits produced by him, and on all spirits (including denatured spirits) now or hereafter in transit to, or received at, his distillery; and

   (b) as a bonded warehouseman, pay, or cause to be paid, to the United States, all taxes imposed by law now or hereafter in force, plus penalties if any, and interest, on all distilled spirits (including denatured spirits) now or hereafter deposited in said bonded warehouse, before withdrawal thereof from said bonded warehouse (except as otherwise provided by law), and within 20 years (except as otherwise provided by law) from the date of original entry for deposit in internal revenue bond, and on all distilled spirits (including denatured spirits) now or hereafter in transit thereto or received thereat; and

   (c) as a rectifier, pay, or cause to be paid, to the United States, all taxes imposed by law now or hereafter in force, plus penalties if any, and interest, for which he may become liable; and

   (d) as the proprietor of a bonded wine cellar, pay, or cause to be paid, to the United States, all taxes, including all occupational and rectification taxes, imposed by law now or hereafter in force (plus penalties if any, and interest) for which he may become liable with respect to operation of the said bonded wine cellar, and on all distilled spirits and wine now or hereafter in transit thereto or received thereat, and on all distilled spirits and wine removed therefrom, including wine withdrawn without payment of tax, on notice by the principal, for exportation, or use on vessels or aircraft, or transfer to a foreign-trade zone, and not so exported, used, or transferred, or otherwise lawfully disposed of or accounted for: *Provided,* That this obligation shall not apply to taxes on wine in excess of $100 which have been determined for deferred payment upon removal of the wine from the premises of the said bonded wine cellar or transfer to a taxpaid wine room thereon; and

4. As a distiller, shall not suffer the property, or any part thereof, subject to lien under 26 U.S.C. 5004(b) (1), to be encumbered by mortgage, judgment, or other lien during the time in which he shall carry on such business (except that this condition shall not apply during the term of any indemnity bond given under 26 U.S.C. 5173 (b) (1) (C), or to any judgment or other lien covered by a bond given under 26 U.S.C. 5173 (b) (4)); and

5. Shall, as to all distilled spirits (including denatured spirits) removed from the bonded premises free of tax, faithfully comply with all the requirements of law and regulations pertaining thereto; and

6. Shall, as to all distilled spirits withdrawn from the bonded premises without payment of tax, on application of the proprietor of the bonded premises, for exportation or for use on vessels or aircraft, or for transfer to a foreign-trade zone, or for transfer to a customs bonded warehouse, or for research, development or testing, as authorized by law—

   (a) faithfully comply with all the requirements of law and regulations pertaining thereto; and

   (b) as to the said distilled spirits or any part thereof withdrawn for exportation or for use on vessels or aircraft or for transfer to a foreign-trade zone or for transfer to a customs bonded warehouse, or for research, development, or testing, and not so exported, used or transferred, or otherwise lawfully disposed of or accounted for, pay to the United States the tax imposed thereon by law, now or hereafter in force, together with penalties and interest;

   Then this obligation is to be null and void, but otherwise to remain in full force and effect.

We, the obligors, for ourselves, our heirs, executors, administrators, successors, and assigns, also agree *(a)* that all stipulations, covenants, and agreements of this bond shall extend to and apply to any change in the business address of the premises, the extension or curtailment of such premises, including the buildings thereon, or any part thereof, or in equipment, or any other change which requires the principal to file a new or amended registration, application, or notice, except where the change constitutes a change in the proprietorship of the business, or in the location of the premises, and *(b)* that this bond shall continue in effect whenever operation of the distillery or of the rectifying plant is resumed from time to time following suspension of operations by an alternate proprietor, and *(c)* that this bond shall continue in effect, notwithstanding the exclusion of the bottling-in-bond department, or any part thereof, from time to time for use temporarily in the rectification or bottling of distilled spirits on which the tax has been paid or determined.

And we, the obligors, for ourselves, our heirs, executors, administrators, successors, and assigns, do further covenant and agree that upon the breach of any of the covenants of this bond, the United States may pursue its remedies against the principal or surety independently, or against both jointly, and the said surety hereby waives any right or privilege it may have of requiring, upon notice, or otherwise, that the United States shall first commence action, intervene in any action of any nature whatsoever already commenced, or otherwise exhaust its remedies against the principal.

WITNESS our hands and seals this ___2nd___ day of ___January___ 19__80__

Signed, sealed, and delivered in the presence of—

_____   *Dave Strawman* (SEAL)
President of Farm Fuel Inc.

_____   (SEAL)

_____   *Donald Rubin* (SEAL)
Attorney In Fact   Donald Rubin

(SEAL)

ATF Form 2601 (5110.56) (8-78)

DESCRIPTION OF PREMISES OPERATED

NAME	ADDRESS	REG. NO.	OPERATION FOR WHICH QUALIFIED

REGIONAL REGULATORY ADMINISTRATOR'S APPROVAL	REGION

The foregoing bond, having been executed in due form and in compliance with the applicable law, regulations, and instructions, is approved by me on behalf of the United States.

SIGNATURE OF REGIONAL REGULATORY ADMINISTRATOR, BUREAU OF ALCOHOL, TOBACCO AND FIREARMS	DATE APPROVED

INSTRUCTIONS

1. This bond shall be filed in duplicate with the Regional Regulatory Administrator, Bureau of Alcohol, Tobacco and Firearms, of the region where the premises covered by the bond are located.

2. The name, including the full given name, of each party to the bond shall be given in the heading, and each party shall sign the bond, or the bond may be executed in his name by an empowered attorney-in-fact.

3. In the case of a partnership, the firm name, followed by the names of all its members, shall be given in the heading. In executing the bond, the firm name shall be typed or written, followed by the word "by" and the signatures of all partners, or the signature of any partner authorized to sign the bond for the firm, or the signature of an empowered attorney-in-fact.

4. If the principal is a corporation, the heading shall give the corporate name, the name of the State under the laws of which it is organized, and the location of the principal office. The bond shall be executed in the corporate name, immediately followed by the signature and title of the person authorized to act for the corporation.

5. If the bond is signed by an attorney-in-fact for the principal, or by one of the members for a partnership or association, or by an officer or other person for a corporation, there shall be filed with the bond an authenticated copy of the power of attorney, or a resolution of the board of directors, or an excerpt of the bylaws, or other document, authorizing the person signing the bond to execute it for the principal, unless such authorization has been filed with the Regional Regulatory Administrator, Bureau of Alcohol, Tobacco and Firearms, in which event a statement to that effect shall be attached to the bond.

6. The signature for the surety shall be attested under corporate seal. The signature for the principal, if a corporation, shall also be so attested if the corporation has a corporate seal; if the corporation has no seal, that fact should be stated. Each signature shall be made in the presence of two persons (except where corporate seals are affixed), who shall sign their names as witnesses.

7. A bond may be given with corporate surety authorized to act as surety by the Secretary of the Treasury, or by the deposit of collateral security consisting of bonds or notes of the United States. The Act of July 30, 1947 (Section 15, Title 6, U.S.C.) provides that "the phrase 'bonds or notes of the United States' shall be deemed * * * to mean any public debt obligations of the United States and any bonds, notes, or other obligations which are unconditionally guaranteed as to both interest and principal by the United States."

8. If any alteration or erasure is made in the bond before its execution, there shall be incorporated in the bond a statement to that effect by the principal and surety or sureties; or if any alteration or erasure is made in the bond after its execution, the consent of all parties thereto shall be written in the bond.

9. The penal sum named in the bond shall be in accordance with 27 CFR Part 201.

10. If the bond is approved, a copy shall be returned to the principal.

11. All correspondence about the filing of this bond or any subsequent action affecting this bond should be addressed to the Regional Regulatory Administrator, Bureau of Alcohol, Tobacco and Firearms with whom the bond is filed.

ATF FORM 2601 (5110.56) (8-78)

FUEL FROM FARMS

www.KnowledgePublications.com

The Universal Insurance Company
Golden, Colorado
Power of Attorney

KNOW ALL MEN BY THESE PRESENTS:

That UNIVERSAL INSURANCE COMPANY, a Colorado Corporation, does hereby appoint DONALD RUBIN, JOHN SMITHSON, or ANDY ANDERSON - GOLDEN, COLORADO its true and lawful Attorney(s)-in-Fact, with full authority to execute on its behalf fidelity and surety bonds or undertakings and other documents of a similar character issued in the course of its business, and to bind the respective company thereby, in amounts or penalties not exceeding the sum of ONE HUNDRED THOUSAND AND NO/100 Dollars ($100,000.00).
EXCEPT NO AUTHORITY IS GRANTED FOR:

1. Bid or proposal bonds where estimated contract price exceeds the amount stated herein.
2. Open Penalty bonds.
3. Bonds where Attorney(s)-in-Fact appear as a party at interest.

IN WITNESS WHEREOF, UNIVERSAL INSURANCE COMPANY of Colorado, Inc. has executed, and attested these presents
This __2__ day of __Jan__, 19__80__

_____   _____
Horlan Winsom, Asst. Secretary   Derek Mott, President

AUTHORITY FOR POWER OF ATTORNEY

That UNIVERSAL INSURANCE COMPANY, a Colorado Corporation, in pursuance of authority granted by that certain resolution adopted by their respective Board of Directors on the 1st day of March, 1976 and of which the following is a true, full, and complete copy:

"RESOLVED, That the President, any Vice-President, or any Secretary of each of this Company be and are hereby authorized and empowered to make, execute, and deliver in behalf of this Company unto such person or persons residing within the United States of America, as they may select, its Power of Attorney constituting and appointing each such person its Attorney-in-Fact, with full power and authority to make, execute and deliver, for it, in its name and in its behalf, as surety, any particular bond or undertaking that may be required in the specified territory, under such limitations and restrictions, both as to nature of such bonds or undertaking and as to limits of liability to be undertaken by these Companies, as said Officers may deem proper; the nature of such bonds or undertakings and the limits of liability to which such Powers of Attorney may be restricted, to be in each instance specified in such Power of Attorney.

RESOLVED, That any and all Attorneys-in-Fact and Officers of the Company, including Assistant Secretaries, whether or not the Secretary is absent, to and are hereby authorized and empowered to certify or verify copies of the By-Laws of this Company as well as any resolution of the Directors, contracts of indemnity, and all other writings obligatory in the nature thereof, or with regard to the powers of any of the officers of this Company or of Attorneys-in-Fact. RESOLVED, That the signature of any of the persons described in the foregoing resolution may be facsimile signatures as fixed or reproduced by any form of typing, printing, or other reproduction of the names of the persons hereinabove authorized."

CERTIFICATION OF POWER ATTORNEY

I, Horlan Winsom, Asst. Secretary of UNIVERSAL INSURANCE COMPANY, do hereby certify that the foregoing Resolution of the Board of Directors of this Corporation, and the Power Attorney issued pursuant thereto are true and correct and are still in full force and effect.

IN WITNESS WHEREOF, I have hereunto set my hand and affixed the facsimile seal of the Corporation this ___2___ day of ___Jan.___, 19_80_

_____
Horlan Winsom, Asst. Secretary

Farm Fuel Inc.
1617 Cole Boulevard
Golden, Colorado 80401

September 15, 1979

U.S. Treasury Department
Bureau of Alcohol, Tobacco, and Firearms
Regional Regulatory Administrator, Southwest Region
Main Tower, Room 345
1200 Main Street
Dallas, Texas 75202

Dear Sir:

I am president of Farm Fuel, Inc. We are working with the Ethanol Department of the State University to develop the capability to produce alcohol. An in-depth plan has been developed by ourselves and our advisors. I will be personally responsible for the operations of the ethanol production units.

Sincerely,

Dave Strawman

Dave Strawman
President

DS/to

Enclosures

Farm Fuel Inc.
1617 Cole Boulevard
Golden, Colorado 80401

U.S. Treasury Department
Bureau of Alcohol, Tobacco, and Firearms
Regional Regulatory Administrator, Southwest Region
Main Tower, Room 345
1200 Main Street
Dallas, Texas 75202

Subject: Application for a Permit to Operate an Experimental Distilled Spirits Plant for Two Years

A. <u>Nature of Operation</u>: Farm Fuel, Inc. wishes to apply for a permit to operate a farm-sized experimental Distilled Spirits Plant (DSP) for experimentation and demonstration purposes for two to six months. At the end of this period we wish to be producing farm-sized DSPs for resale. The applicant is a corporation, incorporated under the laws of the State of Colorado.

B. <u>How Much Production</u>: Each of our DSPs will produce approximately 25 gal/hr, 600 gal/day, and up to 210,000 gal/yr.

C. <u>Purpose More Specifically</u>: Farm Fuel, Inc. would set up one farm-sized DSP at the manufacturing facility at 1617 Cole Boulevard in Golden, Colorado, for experimentation and demonstration purposes. Our goal is to manufacture and sell farm-sized plants to farmers so they may become energy-independent. This will also help the economy of Golden, the economy of Colorado, and the United States in helping our balance-of-trade deficit.

D. <u>Corporate Authorization</u>:   The President and the Secretary have each been authorized by the Board of Directors to sign applications and other documents required by the Bureau of Alcohol, Tobacco, and Firearms.

E. <u>Plant Security</u>: The entire distilled spirits plant as described herein will be bonded. It will be that land located in the eastern half of the southwest quarter of Section 9, Mean Township, R9W, 10PM, located in Jefferson County, Colorado. The tract of land is completely surrounded by a substantially constructed chain link fence of 9-gauge wire mesh. The fence is 6 feet high and is superimposed at the top by three strands of barbed wire. The entrance gate is equipped for locking. On this land, there will be a single 1-story building which will be of brick construction with a concrete floor and a flat tar and gravel roof over 1-inch wood sheathing, supported on 2-inch by 12-inch joints on 16-inch centers. The entrance door, which will be equipped for locking from the outside, will be located at the center of the southern end of the building. It will be an overhead steel door 15 feet wide and 12 feet high. On either side of this door will be two windows with sills 5 feet from the base of the building. Each of these windows will be 4 feet wide by 8 feet high and each will be protected by securely affixed wire mesh screen of 6-gauge steel, having 1 1/2-inch mesh. At the opposite end of the building there will be another overhead steel door, identical to the entrance door except that it will be equipped for locking from the inside. A third door, also equipped for locking from the inside, will be located on the east side of the building, 50 feet from its northern end. This door will be a 2-inch thick, metal-clad door, 3 feet wide by 7 feet high.

The above described building will be located on the bonded premises as follows: Beginning at a point 10 feet east of the western boundary of the bonded premises and 52 1/2 feet north of the southern boundary, proceed due east for 125 feet, then due north for 200 feet, then due west for 125 feet and, finally, due south for 200 feet to the point of beginning.

The southern quarter of the west side of the building will be partitioned into three offices, as follows:
> The southernmost office will be 200 feet long (in line with the DSP building) by 15 feet wide. The room immediately north of that room will be used for a government office, and it will be 16 feet long by 15 feet wide. The third room will be 14 feet long by 15 feet wide. Each of these three rooms will be well-lighted by fluorescent lighting and will have an entrance on its east wall equipped for locking. In addition, each room will have a window on its east wall affording light as well as a view of the activities within the plant. Each room will be ventilated through the air conditioning and heating system which includes vents for return of stale air through the system. A desk, two chairs, and a filing cabinet will be provided in the government office, and additional furniture will be furnished if needed. Lavatory and toilet facilities will be located immediately north of the three rooms described above.

F. Description of Equipment: On the outside of DSP building, at a site approximately 20 feet from its northern end and 20 feet from its eastern side, will be the beer still and the rectifying column. These are described as follows: Beer Still No. 190, manufactured by I. Lovehooch and Co., Distillate, Ohio, is a 12-inch column still having 24 plates. Rectifying Column No. 200, manufactured by Samson and Co., is also a 12-inch column with 24 plates.

Other Equipment	Capacity
Mash tub and cooker No. 1	1,500 gallons
Fermenter No. 1	2,000 gallons
High wine tank No. 1	5,000 gallons
High wine tank No. 2	4,800 gallons
Denaturing tank No. 1	3,000 gallons

All of the above tanks, except the mash tub and cooker and the fermenter, will be equipped with accurately calibrated sight glasses.

An accurately adjusted, tamper-proof, White Mule electronic meter will be sealed into the draw-off line between the rectifying column and the manifold connecting to the two high wine tanks, and a drip sampling device will also be installed on that line, downstream from the meter. All of the above described "other" equipment will be installed inside the DSP building along the east wall, with the two denaturing tanks 20 feet south of the side door and the remainder of the equipment north of that door.

G. Process Description: Qualification bond will be in the maximum penal sum. We expect to produce a maximum of 600 proof gallons of alcohol per day (24 hours).

Statement of Process—Fuel-grade Fermentation Alcohol

1. Mashing
   a. Place 312 gallons of water into the mash tub.
   b. Slowly add 52 bushels of ground corn, ground wheat, or a mix of these two grains in any proportion found to be expedient. Begin stirring the mixture at the same time and heat the mixture to a temperature of 100°F.
   c. As needed, add either concentrated sulphuric acid or a slurry mix of hydrolated lime, in quantities sufficient to adjust the pH of the mash to between 6.0 and 7.5.
   d. Prepare a slurry composed of 2.6 pounds of TAKA-THERM and 4.5 gallons of water, warmed to 100°F.
   e. Add the TAKA-THERM slurry to the mash while continuing to stir it, and bring the temperature up to 203°F for 30 minutes.
   f. While continuing with the stirring, reduce the temperature of the mash to 194°F.
   g. Hold the temperature at 194°F for 2 hours or until all of the starch has been converted to dextrin, stirring all the while.
   h. Cool the mash to 130°F while continuing to stir it.
   i. By slowly adding concentrated sulphuric acid, adjust the pH of the mash to about 4.0.
   j. Add 6.5 pounds of DIAZYME L-100D, continue to stir, and hold the temperature of the mash at 130°F for 30 minutes.
   k. While the mash continues to be stirred, add sufficient cold water to reduce its temperature to 85°F.
   l. Using a water slurry of hydrated lime, adjust the pH of the mash to 5.5.

2. Fermentation

   a. Stir 0.65 pound of sugar into 2.6 gallons of water at 85°F and add 1.3 pounds of bakers' yeast. Adjust pH to between 6.5 to 7.
   b. As soon as the yeast mixture becomes active, pump the mash from the mash tub to the fermenting tank, and add the yeast mixture to it while pumping.
   c. Let the mash rest in the fermenting tank until fermentation is complete.

3. Distillation

   a. Through a strainer, pump the mash from the fermenting tank into the beer still at plate 22 and inject steam into the bottom of the still. Maintain temperature at the top of the beer still at 175°F.
   b. Vapor will pass through the draw-off line of the beer still at the top of the still and pass through this pipeline into the rectifying column at plate 23.
   c. Alcohol vapors will pass through the top draw-off line of the rectifying column, through the condenser, thence, in liquid form, through the trybox. At that point the alcohol may either be refluxed back to the rectifying column or directed to one of the tanks. From the tanks the alcohol may either be recycled through the distilling system or drawn off pursuant to a production gauge.

All alcohol produced, except for small quantities to be used for laboratory testing, will be completely denatured by entering gasoline through a Y connection converging with the alcohol draw-off line of the two tanks. An A.O. Don proportioning device will be used to control the flow of gasoline so that the resulting completely denatured alcohol will contain 5 gallons of gasoline to each 100 gallons of alcohol.

H. Title: Title to the bonded premises is held in fee simple by the Denver Industrial Complex, Inc. There is an outstanding mortgage on the land and building in the amount of $65,000, held by the First Bank of Kansas, Carbon Grain, Kansas 10020. See attached letter from the Denver Industrial Complex, Colorado.

I. Records: Complete and accurate records will be kept of all runs as to all materials, yeast, corn, denaturant, amount of water, and energy used and yield of alcohol. Accurate records of disposition will also be kept and will be open to inspection.

J. Other Information: No bonded warehousing operations will be conducted. No rectifying operations will be conducted. No bottling operations will be conducted. In the open area in the northeast corner of the building, convenient to the fermenting tank, there will be a press for removing the liquids from the solids to be strained from the mash, before it enters the beer still. These solids will be sold as cattle feed to either or both of two nearby feedlots. The business of negotiating sales of such feedstocks and of keeping the necessary records of sales transactions will be conducted in the larger of the two company offices.

Dave Strawman, President

The Denver Industrial Complex

U.S. Treasury Department
Bureau of Alcohol, Tobacco, and Firearms
Regional Regulatory Administrator, Southwest Region
Main Tower, Room 345
1200 Main Street
Dallas, Texas 75202

The Denver Industrial Complex, a partnership, owner of the property described in Exhibit "A" and "B" attached does hereby authorize Farm Fuel, Inc. to set up and operate grain alcohol distilling equipment on the described property.

The Denver Industrial Complex further authorizes inspectors of the Federal Bureau of Alcohol, Tobacco, and Firearms to enter described property during regular working hours to inspect equipment and records of Farm Fuel, Inc.

The Denver Industrial Complex

J. Arnold Hammer
J. Arnold Hammer

Dusty Rhodes
Dusty Rhodes

Dan Scherr
Dan Scherr

SAVE HARMLESS AGREEMENT

Because use of the information, instructions and materials discussed and shown in this book, document, electronic publication or other form of media is beyond our control, the purchaser or user agrees, without reservation to save Knowledge Publications Corporation, its agents, distributors, resellers, consultants, promoters, advisers or employees harmless from any and all claims, demands, actions, debts, liabilities, judgments, costs and attorney's fees arising out of, claimed on account of, or in any manner predicated upon loss of or damage to property, and injuries to or the death of any and all persons whatsoever, occurring in connection with or in any way incidental to or arising out of the purchase, sale, viewing, transfer of information and/or use of any and all property or information or in any manner caused or contributed to by the purchaser or the user or the viewer, their agents, servants, pets or employees, while in, upon, or about the sale or viewing or transfer of knowledge or use site on which the property is sold, offered for sale, or where the property or materials are used or viewed, discussed, communicated, or while going to or departing from such areas.

Laboratory work, scientific experiment, working with hydrogen, high temperatures, combustion gases as well as general chemistry with acids, bases and reactions and/or pressure vessels can be EXTREMELY DANGEROUS to use and possess or to be in the general vicinity of. To experiment with such methods and materials should be done ONLY by qualified and knowledgeable persons well versed and equipped to fabricate, handle, use and or store such materials. Inexperienced persons should first enlist the help of an experienced chemist, scientist or engineer before any activity thereof with such chemicals, methods and knowledge discussed in this media and other material distributed by KnowledgePublications Corporation or its agents. Be sure you know the laws, regulations and codes, local, county, state and federal, regarding the experimentation, construction and or use and storage of any equipment and or chemicals BEFORE you start. Safety must be practiced at all times. Users accept full responsibility and any and all liabilities associated in any way with the purchase and or use and viewing and communications of knowledge, information, methods and materials in this media.